Business Models for Renewable Energy in the Built Environment

T0186829

This book provides insight to policy makers and market actors regarding the way that new and innovative business models can stimulate the deployment of renewable energy technologies and energy efficiency measures in the built environment.

The book analyses ten business models in three categories, including amongst others, different types of Energy Service Companies, developing properties certified with a 'green' building label, building owners profiting from rent increases after implementing energy efficiency measures, Property Assessed Clean Energy financing, on-bill financing, and leasing of renewable energy equipment, as well as their organisational and financial structure, the existing market and policy context, and an analysis of strengths, weaknesses, opportunities and threats.

After looking at all of these elements, the book concludes with recommendations for policy makers and other market actors on how to encourage and accelerate the deployment of renewable energy technologies in the built environment.

The basis for this publication was a project initiated and funded by the International Energy Agency's Implementing Agreement for Renewable Energy Technology Deployment (IEA-RETD). Working under the legal framework of the International Energy Agency, the IEA-RETD was officially launched in September 2005 with five founding members. Current members of the IEA-RETD are Canada, Denmark, France, Germany, Ireland, Japan, Netherlands, Norway, and the United Kingdom. The IEA-RETD's mandate is to address cross-cutting issues influencing further deployment of renewable energy and to act as a vehicle to accelerate the market introduction and deployment of renewable energy technologies. More information on the IEA-RETD can be found on the organisation's webpage at: http://iea-retd.org/

About IEA-RETD

RETD stands for 'Renewable Energy Technology Deployment'. IEA-RETD is a policy-focused, technology cross-cutting platform that brings together the experience and best practices of some of the world's leading countries in renewable energy with the expertise of renowned consulting firms and academia. IEA-RETD is a so-called Implementing Agreement, i.e. a platform where a number of countries cooperate under the framework of the International Energy Agency (IEA).

The mission of IEA-RETD is to accelerate the large-scale deployment of renewable energies (RE). This is achieved by providing information and recommendations on RE technology cross-cutting issues to policy makers and other stakeholders. To this end, IEA-RETD commissions annually 5–7 studies performed by consultancies and academia. The reports and handbooks are publicly and freely available on the IEA-RETD's website at www.iea-retd.org. In addition, IEA-RETD organises at least two workshops per year and presents at national and international events. IEA-RETD has been in existence since 2005 and has currently 9 member countries (Canada, Denmark, France, Germany, Ireland, Japan, Netherlands, Norway and the UK).

Business Models for Renewable Energy in the Built Environment

IEA-RETD

from Routledge

IEA - RENEWABLE **ENERGY** TECHNOLOGY **DEPLOYMENT**

First published 2013
by Routledge
2 Park Square, Milton Park, Abingdon, Oxon, OX14 4RN

Simultaneously published in the USA and Canada
by Routledge
711 Third Avenue, New York, NY 10017

Routledge is an imprint of the Taylor & Francis Group, an informa business

British Library Cataloguing in Publication Data
A catalogue record for this book is available from the British Library

Library of Congress Cataloging-in-Publication Data
  Business models for renewable energy in the built environment /
  IEA-RETD.
    p. cm.
  Includes index.
  1. Real estate development—Environmental aspects.
  2. Residential real estate—Environmental aspects.
  3. Sustainable buildings—Economic aspects. 4. Sustainable
  construction—Economic aspects. 5. Renewable energy sources—
  Economic aspects. I. International Energy Agency.
  HD1390.B87 2013
  333.79'413—dc23   2012023537

ISBN13: 978-0-415-63868-5 (pbk)
ISBN13: 978-0-203-08317-8 (ebk)

Typeset in Sabon
by Keystroke, Station Road, Codsall, Wolverhampton

Printed in Spain by GraphyCems

Contents

List of illustrations

Figures

Tables

Boxes

Acknowledgements

This book presents the outcomes of the project 'Business Models for Renewable Energy in the Built Environment' (RE-BIZZ), initiated and funded by the IEA Implementing Agreement for Renewable Energy Technology Deployment (IEA-RETD). It was first published as a project report in November 2011. It was updated in April 2012 with four additional case studies. The contact person for this publication at ECN is Laura Würtenberger (wuertenberger@ecn.nl).

The authors would like to thank the Project Steering Group represented by Michael Paunescu (Natural Resources Canada), Kjell-Olav Skjølsvik (Enova Norway), Milou Beerepoot (IEA Secretariat), Walt Patterson (Chatham House), David de Jager (Ecofys, IEA-RETD Operating Agent), Kristian Petrick (IEA-RETD Operating Agent) for their review and guidance during the course of the project. A special thank you goes to Kristian Petrick, who was always available for feedback and valuable discussions. The authors would also like to thank Ron van der Steen from Financial Consult Nederland for his helpful feedback and review, and Sytze Dijkstra at ECN for co-reading the report.

Summary

The project 'Business Models for Renewable Energy in the Built Environment' (RE-BIZZ) aims to provide policy makers and other market actors with insight into the way new and innovative business models can stimulate the deployment of renewable energy technologies (RET) and energy efficiency (EE) measures in the built environment.

Today, various barriers prevent an increased deployment of RET in the built environment including:

- market and social barriers: price distortion through externalities, low priority of energy issues, split incentives, etc.
- information failures: lack of awareness, knowledge and competence
- regulatory barriers: restrictive procurement rules, cumbersome building permitting processes
- financial barriers: low (or no) returns on investment, high up-front costs, lack of access to capital etc.

For the scope of this study, a business model was defined as *'a strategy to invest in RET (and EE measures), which creates value and leads to an increased penetration of RET in the built environment'*. Successful business models represent approaches in which the financing and implementation of RET or EE in buildings is organised in such a way that certain barriers for the deployment of RET are overcome. Based on the main drivers for value creation, business models for RET in the built environment can be grouped in three categories, in which overall 10 business models were analysed:

- *Product service systems/energy service companies (ESCOs)*

1 Energy supply contracting (ESC): an energy service company (ESCO) supplies useful energy, such as electricity, hot water or steam to a building owner (as opposed to final energy such as pellets or natural gas in a standard utility contract). The ESC model is particularly well suited for generating electricity and heat from RET.

2 Energy performance contracting (EPC): an ESCO guarantees energy cost savings in comparison to a historical (or calculated) energy cost baseline. For its services and the savings guarantee the ESCO receives a performance-based remuneration.

3 Integrated energy contracting (IEC): the IEC model is a hybrid of ESC and EPC aiming to combine supply of useful energy, preferably from renewable sources with energy conservation measures in the entire building. The model is currently being piloted in Austria and Germany.

- *Business models based on new revenue models*

4 Making use of a feed-in remuneration scheme: through a feed-in remuneration scheme the producer of renewable energy receives a direct payment per unit of energy produced. A feed-in scheme guarantees access to a predictable and long-term revenue stream, which can serve as a stable basis for a business model.

5 Developing properties certified with a green building label: 'green' building certification systems assess a building's performance according to environmental and wider sustainability criteria. In this business model a property developer or architect designs and builds buildings certified according to a voluntary 'green' certification scheme, expecting to realise a sales price premium compared to conventional buildings.

6 Building owner profiting from rent increases after the implementation of energy efficiency measures: building owners who do not occupy a building themselves or housing corporations can profit from additional revenue opportunities after undertaking investments in RET and EE measures if they are allowed to charge higher rent from their tenants after the renovation.

- *Business models based on new financing schemes*

7 Property Assessed Clean Energy (PACE) financing: PACE financing is a mechanism set up by a municipal government by which property owners finance RET and EE measures via an additional tax assessment on their property.[1] The property owners repay the 'assessment' over a period of 15 to 20 years through an increase in their property tax bills. When the property changes ownership, the remaining debt is transferred with the property to the new owner.

8 On-bill financing: utilities provide financing (i.e. a loan) for RET and EE measures. The building owners (or building users) repay the loans via a surcharge on their utility bills.

9 Leasing of renewable energy equipment: leasing enables a building owner to use a renewable energy installation without having to buy it. The installation is owned or invested in by another party, usually a financial

institution such as a bank. Leasing can be a central component of the business model of an ESCO or of a company that introduces a new technology to the market.

10 Business models based on energy saving obligations: energy saving obligations are a policy instrument that obliges energy companies to realise energy savings at the level of end-users. It stimulates business models based on financial incentives offered by energy suppliers to building owners, renters or energy service companies.

The analysis of the business models included an analysis of the organisational and financial structure, the existing market and policy context and an analysis of strengths, weaknesses, opportunities and threats (SWOT). Some of the analysed business models are specific to a certain market segment (e.g. new versus existing, owner-occupied versus rented, residential versus commercial buildings), whereas others can easily be generalised. Practical experience with the models varies among countries.

Strong role of policy makers required

The study demonstrates that business models can play an important role in increasing the deployment of RET in the built environment. They provide opportunities for building owners, e.g. facilitating access to capital, financing of up-front costs, outsourcing of technical and economic risks, and offering further energy related services. In many cases business models require only a supporting role by government, e.g. through changes of legislation. However, business models alone will not lead to a significantly increased deployment of RET. The analysed business models generally only lead to a deployment of cost-effective technologies because they are unable to improve the returns on investment of RET and EE measures by themselves. Moreover, business models cannot address all barriers, e.g. no business model addresses the barrier of 'low priority of energy issues', which keeps building owners from taking action. This implies that a strong role of policy makers is still required.

In which market segments can the business models be applied?

The built environment is a complex sector where barriers for an increased deployment of RET differ *in existing and new, large commercial, residential and public buildings* among market segments. The results show that ESCO models can address the barriers of high up-front costs and access to capital. In *small residential and commercial buildings* this can be achieved by PACE or on-bill financing. These business models make a life-cycle approach possible where building owners can spread the investment costs across the project lifetime. For business models to work in *rented buildings*, the split incentives barrier must be addressed. One way of doing this in regulated rental sectors, especially the social housing sector, involves a change in legislation, allowing building owners to pass on the cost of the investment to the tenant through a rent increase. To cushion the social effects of the measure, the benefits of energy savings should be higher than the rent increase for the tenants. Business models have the advantage that they can work well for *existing buildings* whereas building codes/obligations so far tend to be limited to new buildings and substantial renovations.

Business models for non-cost-effective technologies

Today, there are already many cost-effective opportunities for a deployment of RET and EE measures (e.g. insulation of buildings, solar water heating in sunny climates), although cost-effectiveness largely depends on the background situation. For technologies that are not (yet) cost-effective, business cases may be based on supporting policy measures such as feed-in remuneration schemes. 'Green' certification of buildings can stimulate investments in RET even when they are not cost-effective. However, because such certification is voluntary, it typically only works in niche markets.

Energy saving obligations are introduced by governments to stimulate EE measures and energy services through the participation of energy suppliers. In practice, this policy measure promotes, for example, the role of ESCOs and on-bill financing but originally it only focused on EE. The scope of energy saving obligations could be broadened to include RET in the built environment.

Recommendations for policy makers

- Policy makers should first analyse the *cost-effectiveness* of RET/EE measures in different market segments of the built environment within their jurisdiction.
- To support cost-effective RET in *existing and new large commercial, residential and public buildings* policy makers can stimulate ESCO models, e.g. by supporting market facilitators, facilitating access to finance and changing procurement rules for public buildings.
- To support cost-effective RET in *smaller residential and commercial buildings*, policy makers can stimulate business models such as on-bill financing or PACE financing, e.g. by:

  - deciding on the most promising model based on a stakeholder analysis (which actors have an interest in RET, the ability to offer access to capital, the technical capacity and access to the decision makers)

- mandating or strongly incentivising utilities, e.g. through energy savings obligations to take an active role
- clarifying outstanding legal issues, e.g. on linking liabilities to a property.

- To address split incentives in *rented buildings*, depending on how their rental market is regulated, policy makers may change rental legislation to make rent increases possible after RET or EE investments.

Recommendations for building owners

Public building owners play a special role, as they can serve as a role model and a means to drive the implementation of government targets for RET deployment and energy efficiency in the built environment. Governments can be proactive in applying suitable business models. Public building owners can, for example:

- apply certification with voluntary 'green' building labels to new buildings and during substantial renovation of existing facilities, and;
- directly support ESCO business models by using these models in the public building stock. This may require a change in public procurement rules.

This provides a unique opportunity for local governments to become active in increasing the deployment of RET in the built environment.

The analysis also shows that often business models are most successful when they are based on *partnerships* between actors with complementary expertise and resources, e.g. regarding access to capital, technical expertise and access to the clients/building owners.

Notes

1 'Tax assessments' are comparable to loans as the property owner pays off its debt in instalments over a period of various years.

Chapter 1

Introduction: RET in the built environment

Background

The Implementing Agreement on Renewable Energy Technology Development of the International Energy Agency (IEA-RETD) has the objective to support a significantly higher utilisation of renewable energy technologies (RET) by encouraging more rapid and efficient deployment of these technologies. RET are increasingly recognised for their potential role within a portfolio of low-carbon and cost-competitive energy technologies capable of responding to the dual challenge of climate change and energy security. Moreover, RET have the potential to reduce environmental pollution caused by fossil fuel-based energy sources.

The building sector presents a large opportunity for reducing CO_2 emissions in a cost-effective manner. About 40 per cent of final energy consumption takes place in existing buildings, and buildings account for about 24 per cent of global CO_2 emissions.[1] At the same time, the building sector offers some of the largest potentials for reducing greenhouse gas (GHG) emissions at negative costs. The Intergovernmental Panel on Climate Change (IPCC, 2007) estimates that globally about 30 per cent of the business-as-usual CO_2 emissions in buildings projected for 2020 could be mitigated in a cost-effective way. There is a large potential for meeting the energy demand of buildings by means of district heating and cooling schemes or through the direct use of RET in buildings (IPCC, 2011).

However, as illustrated in previous studies by the IEA (IEA, 2007, 2008a, 2010; IEA-RETD, 2007) and other organisations (e.g. WBCSD, 2010; Wuppertal Institute, 2010; European Commission, 2010/11) various barriers prevent the accelerated uptake of RET and energy efficiency measures in the built environment. New and innovative business models may help to exploit the potential of a sustainable energy in the built environment by addressing one or more of these barriers.

The IEA-RETD therefore commissioned the project 'Business models for Renewable Energy in the Built Environment (RE-BIZZ)' to gain insights into the way new business models and/or policy measures can stimulate the deployment of renewable energy technology (RET) and energy efficiency (EE) in the built environment. The project aims at providing recommendations to both policy makers and market actors. This book presents the work undertaken within this project.

Scope of the report

Technological focus, market segments and country focus

The study focuses on business models for increasing the deployment of RET in the built environment. Where necessary, the report also addresses energy efficiency measures and how energy efficiency measures relate to the deployment of renewable energy, as energy efficiency plays an important role in reducing energy use in buildings. In addition, many existing studies, for example on barriers for reducing GHG emissions from buildings, focus on energy efficiency. Previous research commissioned by the IEA-RETD (IEA-RETD, 2010) suggests that the lessons from the promotion of residential energy efficiency may largely be transferred to programmes promoting the residential use of renewable energy.

The analysis covers renewable electricity, and heating and cooling. The following renewable energy technologies in buildings fall under the scope of the study:

- Solar PV (see Figure 1.1)
- Solar thermal for water and space heating (solar boilers) (see Figure 1.2)
- Small-scale wind turbines on the roofs of buildings for electricity generation
- Biomass heating (e.g. wood pellets)
- Heat pumps and small-scale district heating/combined heat and power (CHP) plants based on renewable energy (e.g. when installed by a property developer on a large housing or business complex)
- Heat and cold storage systems
- Micro-CHP systems may be included because, although they are not a RET, a micro-CHP system is generally more efficient than traditional electricity and heat production, and may be based on renewable energy in the future.

EE measures are not an explicit focus of the report. However, where the analysis does refer to EE measures, these could include the following:

- Insulation (wall, roof, floor, window, heating and water pipes, crack sealing)
- Low temperature room heating
- Heating boiler controls
- Heat recovery systems (ventilation system, shower) (see Figure 1.3)
- Other (water saving shower heads, weatherstrips etc.).

Figure 1.1 New houses with roof-mounted solar PV systems in the Netherlands (photo: ECN)

Figure 1.2 Solar thermal water heating system at the University of Michigan (photo: Clean Energy Resource Teams)

The study distinguishes between the following market segments: new versus existing buildings, owner-occupied versus rented, and commercial versus residential (if needed further split into multi-family dwellings, de/attached homes and stand-alone houses). Within the segment of commercial buildings, where required, the specific role of public building owners is addressed. Deploying RET and EE measures in existing buildings is especially important as in many OECD countries a large part of the housing stock has been constructed before comprehensive building related energy regulations were put in place (see Figure 1.4).

Figure 1.3 Colourful wind cowls which provide passive ventilation with heat recovery at the UK eco-village BedZED, a residential and workspace development in the London borough of Sutton (Photo: telex4)

Some parts of the study include country specific explanations. Case studies from a country or region are used to illustrate the business models. In addition, the business models are put into the context of the country specific regulatory environment. Where this is the case, the IEA-RETD member countries, i.e. Canada, Denmark, France, Germany, Ireland, Japan, Netherlands, Norway, United Kingdom, are examined. The study may also refer to other countries and country situations which could be potentially interesting in the long term for the business models evaluated such as, but not limited to, China and the United States.

How to define business models for RET in the built environment

Research on business models originated during the rise of e-commerce and the development of other internet-based companies in the 1990s and early 2000s. Since then, business models have become an increasingly popular

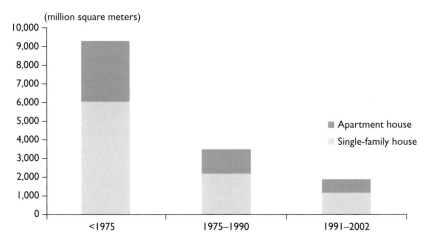

Figure 1.4 Age of the housing stock in Europe. About 50 per cent of building space was built before 1975 (WBCSD, 2010)

concept in management theory and practice. Today, the concept is being applied to an ever wider range of sectors and topics (Wüstenhagen & Boehnke, 2008; Osterwalder, 2004).

A large number of studies on the theory of business models exist, but so far there is no generally accepted definition of what a business model is, although the definitions generally state that it describes how a business creates value (Osterwalder, 2004; Osterwalder et al., 2005; Porter, 2001; Schafer et al., 2001). The approach for value creation can then be split into different aspects, including, for example, the strategic objective and value proposition, sources of revenue, critical success factors, core competencies, customer segments, sales channels (Weill and Vitale, 2001, see Table 1.1) and key activities and resources.

Other definitions are simpler, e.g. defining a business model as 'the method of doing business by which a company can sustain itself, that is, generate revenue' (Rappa, 2001).

Based on these considerations, we recognise the following distinction between a business case and a business model:

- a *business case* captures the logic and reasoning for initiating an activity, such as an investment in RET in the built environment. The reasoning includes a financial calculation demonstrating the profitability of the planned investment.

Table 1.1 Elements of a business model according to Weill and Vitale (2001) (adapted from Osterwalder, 2004)

Element of the business model	Description
Strategic objective and value proposition	Overall view of the target customers, product and service offering and the unique market position that the firm aims to achieve; defines the choices and trade-off inherent in the firms strategy and operation.
Sources of revenue	A firm should have a realistic view on revenue sources.
Critical success factors	In order to make business models successful, the critical structures have to be identified, e.g. by questioning industry assumptions and patterns of established business models.
Core competencies	In-house competencies that a firm needs to possess to be successful.
Customer segments	Important to understand which customer segments a firm targets and by what the specific value proposition for each customer segment is.
Sales channels	The way and approach how a firm's products and services are offered. It is considered part of the firm's value proposition by Weill and Vitale (2001).

- a *business model* describes the structure and strategy behind a business case, and includes elements such as value proposition, key activities, key resources, cost structure and revenue streams. The aim of a business model is to help structure an initiative in a way that leads to a positive business case, one that leads to initiating the activity.

For the scope of this study, a business model is defined as:

a strategy to invest in RET and in EE measures, which creates value and leads to an increased penetration of RET and EE measures in the built environment.

Research on business models generally focus on the strategy at a company level. However, for the concrete case of deployment of RET in the built environment, we broaden the definition of a business model to also include strategies of non-corporate actors. The built environment is an exceptionally multifaceted system, including different market segments and market actors. The World Business Council for Sustainable Development (WBCSD) (2009) in its *Roadmap for a Transformation of Energy Use in Buildings* identifies, for example, seven different group of actors in the sector: government authorities; building developers; investors; occupiers; suppliers and manufacturers; architects, engineers, contractors and craftsmen; and utilities. In addition, energy service companies may be involved. At least five of these, i.e. building developers, investors (i.e. building owners), occupiers, energy service companies and utilities may be directly involved in investing in RET in the built environment, and creating value from this investment.

Business models vary from being relatively simple to being complex. More simple models exist when an actor takes advantage of an existing incentive scheme for RET,[2] while more complex models include Energy Service Companies (ESCOs) offering energy services ranging from providing information and advice, identifying potential RE or EE measures, implementing them, and undertaking operation and maintenance services and financing.

Today, various barriers prevent an increased deployment of RET in the built environment (see Chapter 2). Successful business models represent situations in which the financing and implementation of RET or EE in buildings is organised in such a way that certain barriers for realisation of renewable energy are (to some degree) overcome. Financial barriers such as long payback times, (perceived) high costs and access to capital are major barriers for the implementation of RET (see Chapter 2). Therefore the financial structure of the business models is an important element in the description and analysis of business models in this study.

In addition, the regulatory environment plays a crucial role for business models for the increased deployment of RET. Policy interventions address the barriers for an increased deployment, either by direct incentives (e.g.

subsidies or preferential pricing), or by changing the regulatory framework (e.g. minimum technology standards, obligations). Policy interventions involving financial incentives usually *directly* stimulate the financial structure of the business model. Policies in the category of regulatory schemes tend to be *indirectly* beneficial to business models, e.g. by changing the competitiveness versus conventional energy. In practice, business models may depend on multiple policies, including both incentives and favorable regulatory schemes. This study analyses under which regulatory environment business models are viable.

Reading guide and methodology

This book consists of six chapters:

- Chapter 1 presents some background to the project and describes the scope of the report.
- Chapter 2 identifies current barriers to introducing RET/EE measures in the built environment and describes what these barriers imply from a business case perspective.
- Chapter 3 presents categories of business models for sustainable energy in the built environment and introduces the new and innovative models that are analysed in the report.
- Chapter 4 describes and analyses these business models in detail with respect to their potential for supporting an increased deployment of RET in the built environment.
- Chapter 5 presents a synthesis and conclusion of the business model analysis.
- Chapter 6 gives recommendations for policy makers and market actors.

The identification of current barriers for an increased deployment of RET/EE measures in the built environment in Chapter 2 is based on a literature review of recent studies. Barriers are grouped in four categories based on IPCC (2007), UNEP (2007) and IEA (2008a). The categories of business models presented in Chapter 3 are derived based on the taxonomy of business models in existing studies in general, and specifically on categories of business models for environmental services and sustainable energy. Business models can be categorised according to the main drivers for value creation. For environmental services and sustainable energy, three main categories were identified based on Wüstenhagen and Boehnke (2008) and COWI (2008). These categories were confirmed by a review of existing and planned business models confirming that all potential business models can indeed be summarised under these categories.

To select concrete business models for further analysis, information on a wide range of existing and planned business models was collected. In addition, the study considered how existing and planned legislation and other potential drivers for business models such as an increased awareness of climate change may lead to new business models. We also considered how current barriers for the deployment of RET could theoretically be overcome by business strategies, and how certain business cases including RET become more viable if fossil fuel prices continue to rise. Based on a longer list, in collaboration with the Project Steering Group ten business models were selected for further analysis.

Methodology for business model analysis

The analysis of these business models (see Chapter 4) follows the same general template for all business models to ensure comparability. Only for the energy contracting models some parts of the analysis are presented at the general level of energy contracting models, whilst other elements are explained for the specific sub-models in order to avoid lengthy repetitions of information. The template for the business model analysis contains amongst others a description of the organisational and financial structure and an analysis of strengths, weaknesses, opportunities and threats (SWOT). Strengths and weaknesses consider the business model as such, while opportunities and threats look at the conditions for effective implementation and the impact of external developments.

The organisational and financial structure of the business models is illustrated in a schematic representation which highlights the business model's most important elements (see for example Figure 4.3) based on the formalism developed by Weill and Vitale (2001) for so called e-business initiatives. Similar diagrams are used frequently in different contexts, e.g. by Bleyl (2009) for ESCO business models.

The questions for the SWOT analysis are partly based on the Impact Assessment Guidelines by the European Commission (EC, 2009a). The leading question is, how suitable the business model is to contribute to an increased deployment of RET (and EE measures) in the built environment. Specific questions for the analysis of strengths and weaknesses (i.e. effectiveness, efficiency, usefulness?) are:

- Is the business model effective? Does it lead to an increased penetration of RET?
- Which barriers are removed, or to which extent decreased (usefulness)?
- Is the business model realised cost effectively?
- Does it require a lot of time or effort for the person who implements it?
- Are significant transaction costs involved?
- Can the business model be scaled up, or replicated in other countries?
- Are other policy measures needed as supporting measures, e.g. information campaigns?

Questions for the opportunities and threats (how viable and how vulnerable is the business model?) are:

- In which policy context or under what market conditions does this business model work?
- How do changes in the policy context and market environment affect the business model?
- Is the business model sustainable after financial incentives are discontinued?
- Impacts of technology developments? Impacts of developments in the building stock?
- Impacts of fossil fuel prices and feedstock prices?

However, these questions are only meant to give guidance, as it is not possible to give comprehensive answers in the frame of this study. Ideally the SWOT analysis would be based on three different types of information:

- Information from concrete case studies.
- Information from market and evaluation studies, e.g. on the impact of a certain policy instrument on a business model.
- Generalised statements on the effectiveness, efficiency, usefulness and on the viability and vulnerability of the business model.

However, in reality, this depth of information is not available. Depending on the availability of information on specific business models, the SWOT analyses differ in length and level of elaboration.

The business model analysis is complemented by case studies which show concrete example of the business model in a specific context. The business model analysis focuses on generalised concepts. Reality may be more complex than the stylised business models discussed in this report. The case studies give some insight as to the complexity and variations found in real implementation.

Methodology for synthesis, conclusions and recommendations

The synthesis chapter evaluates the business models and puts them into a larger perspective. Thereby the question is addressed how the analysed business models can stimulate an increased deployment of RET in the built environment. The discussion also touches upon additional questions, such as: will business models that are mainly based on voluntary actions be sufficient to increase the energy efficiency and use of RET in buildings? Or are stricter policy measures required, e.g. like the European Building Performance Directive or local solar ordinances? Are these policies, that apply both 'sticks and carrots', sufficient? How can it be assured that the necessary investments can be financed? And how can it be ensured that the rehabilitation of existing building stock is tackled quickly? However, based on the research undertaken for this book, it is not possible to give comprehensive answers to these questions.

Overview tables which illustrate which barriers are addressed by the business models, in which market segments the business models work, and which actors are directly involved form the basis for the comparison and synthesis. The synthesis leads to some general conclusions, which form the basis for recommendations for policy makers and market actors. In addition, the SWOT analyses lead to recommendations for specific business models.

Notes

1 This is the case in most countries of the world, i.e. both globally (UNEP, 2007) and in OECD countries (IEA, 2008a; EC, 2010/2011).
2 Wüstenhagen and Boehnke (2008) for example consider 'intelligent management of available subsidies' a potentially important element of business models for sustainable energy.

Chapter 2

Overcoming barriers for the deployment of renewable energy technologies (RET) in the built environment

Current barriers

As illustrated in previous studies by the IEA (IEA, 2007, 2008a, 2010; IEA-RETD, 2007) and other organisations (e.g. WBCSD, 2010; Wuppertal Institute, 2010; European Commission, 2010/11) various barriers prevent the accelerated uptake of RET and EE measures in the built environment. Most of the existing studies have focused on barriers to increasing energy efficiency in the built environment, while some recent studies specifically address barriers for an increased uptake of renewable heating and cooling (e.g. IEA-RETD, 2007). In most cases, barriers for RET deployment in the built environment do not differ significantly from barriers for energy efficiency measures, as most barriers are specific to the built environment.

For easier conceptualisation barriers are grouped into four categories (based on IPCC (2007), UNEP (2007) and IEA (2008a)): market and social barriers, information failures, regulatory barriers and financial barriers. As this study explicitly takes an investment/business case perspective, technical barriers are reflected mostly in the higher risk of RET as part of the financial barriers. Political barriers are considered to be part of the regulatory barriers, and market and social barriers. Similarly, behavioural barriers are reflected in market and social barriers, and in financial barriers via high discount rates which hinder up-front capital investments.

Market and social barriers

The following barriers relate to the demand side of the market for RET and EE measures in the built environment.

Low priority of energy issues

In many cases, energy costs in buildings are relatively low when compared to other costs for private persons or companies (IEA, 2007). As a consequence there is little incentive to invest in improving the energy performance of the building. Consumers rather tend to invest in upgrades of their buildings for reasons of comfort, aesthetics, reliability, convenience or status. Companies focus their investments on core business assets, whereas investments into the building stock have only a low priority (IEA, 2007).

Price distortion

From a societal perspective energy is too cheap, as externalities such as the costs of natural resource depletion, health impacts from pollution, and climate change are not included in the market price for energy. This implies that consumers and project developers do not receive accurate price signals reflecting the true marginal cost of energy use.

The 'hassle factor'

The benefits from implementing RET or EE measures may be outweighed by the transaction costs and efforts required for gathering information and the perceived inconvenience of installing new equipment in a building which is in use.

Split incentives

'Split incentives' refers to situations where the investor who pays for the up-front costs for RET or EE measures is not the same person who reaps the benefits of lower energy costs. Split incentives occur for example in rental properties when there is little incentive for the building owner to invest if the tenant pays the energy bill or when there is little incentive for the tenant to save energy when the building owner pays the energy bill (see Figure 2.1). Conversely, the tenant may not be interested in an investment into RET either, as he may move out before the end of the payback period.

There may also be other split incentives, e.g. between project developer and building owner/user in new buildings, where there may be no or little benefit for the developer to incorporate RET, if he does not expect to fully recover the higher initial cost from the building owner/user (IEA-RETD, 2007). Other examples are elderly people or people who may move soon,

Who is responsible for the energy bill	Consequences for	
	Building owner	Tenant
Building owner	Incentive to invest	No incentive to save energy
Tenant	No incentive to invest	Incentive to save energy

Figure 2.1 Split incentives between building owners and tenants (source: WBCSD, 2010)

who are not willing to make any more investments in their houses. There are also less incentives to save energy in rented apartments where the heating costs are evenly split.

In addition, there are barriers to the increased uptake of RET and energy efficiency measures on the supply side of the market.

Fragmentation in the building chain

In most countries, the building development chain is complex and fragmented, which inhibits a holistic approach to building design and use, especially for new buildings. Decisions on RET and energy efficiency measures are taken by different actors, including architects, project developers, construction workers or installers, often without coordination and too late in the development process, even though a successful integration of RET and EE measures requires optimising the system as a whole (IPCC, 2007; WBCSD, 2010). Fragmentation in the sector is also an issue for existing buildings, for example when the installer of a new heating system is not able to advise on related insulation measures.

Lacking intrinsic interest by energy companies

Energy providers often have no intrinsic interest in energy savings by their customers. In addition, they generally do not favor small-scale decentralised solutions, which may compete with their own business model.

Small-scale suppliers of RET

Many small-scale renewable heating and cooling technologies are produced by local, small and medium sized enterprises, where production levels have not reached sufficiently high volume to gain economies of scale (IEA-RETD, 2007). In addition, the lack of standardisation of RET at the regional or global level means that companies may face challenges to penetrate markets abroad. Many suppliers therefore remain small and medium enterprises. These suppliers tend to lack the necessary skills to adequately promote RET products.

Information failures

Lack of awareness

There is a general lack of awareness on RET and EE. If viable RET alternatives are unknown, they are not taken into account in building investments.

Lack of information on financing options

There is a lack of adequate information describing financing options available to individuals investing in EE or RET. Even if building owners are willing to implement EE measures or RET, they often find it difficult to obtain not only qualified, but also independent and objective advice from financial experts. Financiers often have no specific knowledge on EE and RET, and thus will not promote financing such projects.

Lack of knowledge and competence by installers

Lacking knowledge and competence of professionals involved in the installation and maintenance of RET limits the diffusion of RET, as it limits the involvement of these professionals and may lead to poor installation of equipment (IEA-RETD, 2007).

Regulatory barriers

Restrictive procurement rules

Procurement rules may pose barriers to the deployment of RET, for example if governments are not permitted to outsource the management of public buildings to private parties.

Cumbersome building permitting processes

Permits for the installation of RET may be difficult to obtain, or this may be a lengthy process.

Financial barriers

Low (or no) returns on investment

Many RET are not yet cost-competitive with traditional energy technologies (see IEA-RETD (2007) for recent cost estimates of renewable heating and cooling technologies and EC (2008) for recent cost estimates of electricity generation from RET). People tend to not invest in renewable energy or energy saving measures if the payback period is too long or even longer than the economic lifetime of the technology and if the investment does not meet their hurdle criteria.

High up-front costs

Many EE measures require a substantial up-front investment, and most RET have a higher up-front capital cost than conventional technologies. This poses a barrier to investment, as decision makers, especially private home owners may not be willing to make large up-front investments. Fuller (2008), for example, describes implicit discount rates in the order of 25 per cent to 75 per cent for investment decisions by private consumers, which substantially increase the hurdle for any up-front investment.[1]

Difficult access to capital

Low income private home owners and small business owners in particular lack internal capital and face difficulties getting access to external capital for financing RET or EE measures.

Higher risk of RET than of conventional technology

EE and RET projects are often considered risky investments, e.g. because of high technology risk or regulatory risk. Higher risks are included in project evaluations by applying a high discount rate or requiring a higher return on investment to compensate for the risk. As a consequence, EE and RE projects frequently become unattractive to investors. Note that the higher risk of RET can also be perceived rather than real risk. Many RET are already quite advanced and, apart from biomass heating, are not exposed to any fuel price risks, e.g. for purchasing oil or gas.

High transaction costs

From the point of view of service companies or financial institutions, investments in EE measures or RET in individual houses are often relatively small. As technology implementation and associated services such as financing and monitoring of energy savings are complex and thus relatively expensive, small-scale measures are unattractive for investment by commercial banks or involvement of ESCOs.

Incomplete mortgage assessment

For a mortgage, credit capacity and risk profile of customers should improve after implementing EE measures or RET if these lower energy costs, as consequently more income is available to serve interest and down payments. However, mortgage criteria generally do not reflect this and financiers are usually not allowed to acknowledge the increased credit capacity.

Barriers from a business case perspective

Not all barriers described above are relevant for all market segments. Table 2.1 provides an overview of which barriers are relevant for which market segments. The importance of the market segments differs widely among countries, e.g. because the general level of house ownership among the population differs. In some countries, e.g. in the Netherlands and Spain, many people own apartments in multi-family buildings whereas in other countries, e.g. in Germany, apartments in multi-family buildings are primarily rented.

Some barriers are related to energy in general or to general characteristics of renewable energy technologies. These include the 'low priority of energy issues', 'price distortion', 'lacking intrinsic interest by energy companies', 'lack of awareness', 'lack of knowledge and competency by installers', 'cumbersome building permitting process', 'low (or no) returns on investment', 'higher risk of RET than of conventional technologies' and 'high up-front costs'.

Other barriers are specific for some market segments:

- The hassle factor is mostly relevant for existing residential buildings, where the owners occupy the building. In new buildings there is no inconvenience related to installing RET, because the installation takes place before building users move in. In commercial buildings or rented multi-family houses, RET are generally installed on the roof or in a separate room with technical equipment. In rented residential buildings, the decision to invest in RET is taken by the owner based on economical considerations. Here, inconvenience for the tenants is not such an important criterion as in owner-occupied buildings.
- Split incentives are mostly an issue for rented buildings and for property developers of new buildings.
- Lack of information about financing options, mortgage assessment and transaction costs are especially relevant for small-scale projects which comprise of only one single-family house, which is either newly built or owner-occupied. Commercial building owners are expected to have more knowledge about financing options, and in larger buildings or property developments transaction costs relative to the size of the investment in equipment are lower.

Successful business models represent situations in which the financing and implementation of RET in buildings are organised in a way that barriers for realisation of renewable energy are – at least to some degree – overcome. A business model is defined as a strategy where the application of RET creates value, thus from a business case perspective in the first instance financial barriers are most relevant. Financial barriers inhibit value creation if an investment is not profitable or if it is not realised due to lacking access to capital or willingness to make up-front investments. Generally, financial viability is the first requirement for a successful business model, except for some cases where non-financial drivers such as increased comfort, energy security or environmental considerations are primary motives for the deployment for RET.

Table 2.1 Barriers and market segments

Barriers	Residential buildings						Commercial and public buildings			
	New buildings		Existing buildings				New buildings		Existing buildings	
	Built by a project developer	Built by the building owner	Owner-occupied		Rented		Built by a project developer	Built by the building owner	Owner-occupied	Rented
			Multi-family houses	Single family houses	Multi-family houses	Single family houses				
Market and social barriers										
Low priority of energy issues			Applicable to all market segments							
Price distortion			Applicable to all market segments							
The 'hassle factor'			X	X						
Split incentives	X				X	X	X			X
Fragmentation in the building chain		X						X		
Little interest by energy companies			Applicable to all market segments							
Small-scale suppliers of RET			Applicable to all market segments							
Information failures										
Lack of awareness			Applicable to all market segments							
Lack of information on financing		X	X	X						
Lack of knowledge by installers			Applicable to all market segments							
Regulatory barriers										
Restrictive procurement rules									X	X
Building permitting process			Applicable to all market segments							
Financial barriers										
Low returns on investment			Applicable to all market segments							
High up-front costs		X	X	X	X	X				
Difficult access to capital		X	X	X	X	X				
Higher risk of RET			Applicable to all market segments							
High transaction costs		X		X						
Incomplete mortgage assessment		X	X	X		X				

This does not mean that non-financial barriers are not important. Calculations of negative abatement costs demonstrate the significance of other barriers than too low rates of return. WBCSD (2009) calculate that there is a potential for investments of US$ 150 billion in building energy efficiency in the US, Japan, Europe, Brazil, China and India, which would have discounted payback times of five years or less and which would reduce the carbon footprint from the buildings sector by 40 per cent compared to a baseline. McKinsey (2009) in their global Marginal Abatement Cost Curve identify a significant global abatement potential of 2500 Mt CO_2 a year at negative costs in the building sector,[2] which includes renewable energy and energy efficiency measures. These data show that there are significant opportunities for RET and energy efficiency measures in the built environment, which are economically viable, but are not realised because of additional financial barriers such as high up-front investment costs, but also because of a variety of non-financial barriers such as split incentives, information barriers and fragmentation in the building chain (WBCSD, 2009; McKinsey, 2009). For the Netherlands, Figure 2.2 shows, based on a sample of 4,700 households, the percentage of households where different energy efficiency measures could be undertaken in a cost-effective manner.

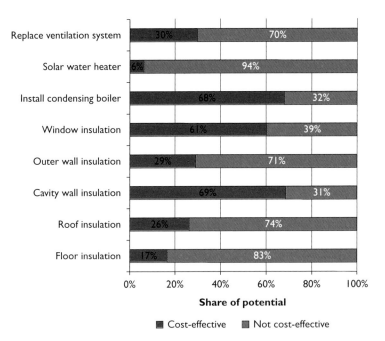

Figure 2.2 Percentage of households where different cost-effective energy efficiency measures can be undertaken based on a sample of 4,700 Dutch households (Tigchelaar et al., 2011)

Often a successful business model that creates a profitable business case for investments into RET in the built environment also addresses some of the non-financial barriers which are important in its market segment. ESCOs, for example, offer a building owner the opportunity to outsource energy related services such as installation, investment, operation and maintenance and fuel purchases. This decreases non-financial barriers such as information and market failures.

The barriers described above illustrate the current situation. However, the barriers are not static, and their importance can change in the future. For example if oil prices and related fossil fuel prices continue to rise, some financial barriers, such as 'low returns on investment' will become less important. The WBCSD (2009) calculations, for example, assume an oil price of 60 USD per barrel. An increased market share of RET may decrease other barriers such as technical risks and information failures. This study focuses on current barriers although the calculation examples in Chapter 5 illustrate how RET become more viable if fossil fuel prices increase.

Notes

1 Assuming a discount rate of 8 per cent as frequently used by policy makers and financial institutions, an intended payback time of 5 years and annual cost savings of $1.000 implies an acceptable up-front investment of $4.000. However, if the consumers' explicit discount rate was 50 per cent rather than 8 per cent, the acceptable up-front investment would decrease to only $1.700.
2 The calculations for McKinsey's global MAC curve are based on a societal perspective assuming a discount rate of 4 per cent. The discount rate for corporate or private investments which would have to be applied for a business case perspective is higher, and is thus expected to lead to a lower abatement potential than the 2,500 Mt CO_2 annually. However, the figure does illustrate that there are significant other barriers inhibiting investments.

Chapter 3

Business models for an increased deployment of RET in the built environment

Categorising business models

New and innovative business models for an increased deployment of RET in the built environment may be categorised according to the main drivers for value creation. Based on Wüstenhagen and Boehnke (2008) and COWI (2008), the following three categories of business models can be distinguished:

 i) Product service systems (PSS)

Product service systems are business models which make use of the delivery of the function of a product combined with a relevant service (COWI, 2008). In the area of energy these are business models offering energy related services beyond the simple sale of energy. Energy service companies (ESCOs) are the most prominent examples of PSS business models in the energy sector.

 ii) Business models based on new and innovative revenue models, or:
iii) on new financing schemes

New and innovative revenue models have been a main driver for new business models in some traditional industries (Wüstenhagen & Boehnke, 2008). For the deployment of RET there are business opportunities in the intelligent use of available government incentives which contribute to revenues. New revenue streams may also emerge from realising the additional value of the intangible climate or environmental benefit of a product, for example of a house with a high rating by a voluntary 'green' building certification scheme. In addition, there are business opportunities in making use of new and innovative financing schemes.

The regulatory environment plays a crucial role for business models for the deployment of RET. Many of the business models that are based on new and innovative revenue models or financing schemes are actually driven by incentive schemes initiated and financed by government. In addition, regulatory schemes such as obligations to deploy RET can be an important driver for investments in RET in the built environment. However, obligations tend to not lead to direct business cases for the market actor who takes the initiative to install RET. But theoretically, an obligation can trigger innovative schemes such as a financing scheme. Such financing schemes emerge, for example, as a consequence of energy saving obligations for utilities.

Alternatively business models can be categorised according to the market segment where they are applicable and according to the main actors involved. The built environment is an exceptionally multifaceted system which includes many different market actors, such as building owners, tenants, government authorities; building developers; financial institutions, suppliers and manufacturers; architects, engineers, contractors, craftsmen and service companies; and utilities. The business model analyses in Chapter 4 describe the applicable market segments and market actors involved.

Table 3.1 shows the business models which are analysed in detail in Chapter 4. These models were chosen because they are considered to have the potential to lead to an increased deployment of RET and/or have the potential to be implemented widely. Additional considerations were:

- If possible, the models should cover all market segments of the built environment.
- If possible, the models should address a wide range of barriers for an increased deployment of RET.
- The selection should cover both very new and innovative models, for which only little experience exists (e.g. integrated energy contracting, PACE financing) as well as models that have been applied widely enough

Table 3.1 List of analysed business models

Business models
Product Service Systems / Energy Contracting models
1 Energy Supply Contracting (ESC)
2 Energy Performance Contracting (EPC)
3 Integrated Energy Contracting (IEC)
Business models based on new revenue models
4 Making use of a feed-in remuneration scheme
5 Developing properties certified with a green building label
6 Building owner profiting from rent increases after the implementation of energy efficiency measures
Business models based on new financing schemes
7 Property Assessed Clean Energy (PACE) financing
8 On-bill financing
9 Leasing of renewable energy equipment
10 Business models based on Energy Saving Obligations

to allow for a comprehensive evaluation of strengths and weaknesses (e.g. on-bill financing, feed-in remuneration schemes).

- The analysis should include energy contracting models as these are frequently discussed as an important market-driven approach for increasing the deployment of RET and EE in the built environment.

Product service system business models

(1) Energy Supply Contracting (ESC) and (2) Energy Performance Contracting (EPC)

Energy service companies (ESCOs) are one of the most prominent examples of product service system business models for sustainable energy. Within the ESCO sector, it is possible to distinguish between two fundamentally different business models which provide either useful energy via (1) energy supply contracting (ESC) or energy savings via (2) energy performance contracting (EPC) to the end-user. Under an energy supply contracting (ESC) model, an energy service company (ESCO) supplies useful energy, such as electricity, heat or steam under a long-term contract to a building owner or building user. The EPC model is based on delivering energy savings compared to a predefined baseline (for more details see Chapter 4). Figure 3.1 depicts typical scopes of services of different ESCO models.

In practice there are also many variations within the ESC and EPC models. Most of these variations related to the range of services delivered under the contracts and to the question how the required investments are financed. In the Anglo-Saxon EPC markets, two EPC models are differentiated mainly with regards to who finances the investment: 'Guaranteed Savings' refers to a service model without financing by the ESCO, whereas 'Shared Savings' includes financing in the ESCO's service package.

(3) Integrated Energy Contracting (IEC)

In addition to the two basic models, a third, innovative approach is being piloted in Austria and Germany, the (3) integrated energy contracting (IEC) model. It is methodologically based on the ESC model and is supplemented by a deemed savings approach for the energy efficiency measures. Compared to standard ESC models, the IEC approach extends the range of services and thus the energy and emissions savings potential to the whole building (see Figure 3.1).

Business models based on new revenue models

(4) Making use of a feed-in remuneration scheme

Feed-in schemes have emerged as one of the most common and successful (in terms of leading to an increased deployment of RET) incentive schemes

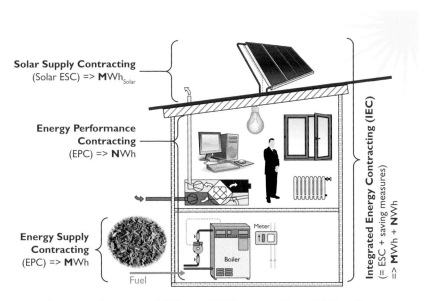

Figure 3.1 Scope of services of different ESCO models Note: NWh refers to energy savings, i.e. avoided MWh. Source: Bleyl (2009)

covering the higher cost of RET versus conventional technologies by compensating the owner of the RET installation with a higher price for the renewable energy. A feed-in remuneration scheme creates opportunities for business cases as it can cover the financial gap between RET and conventional technologies. Feed-in tariffs or feed-in premiums for electricity from renewable sources are the most common. A renewable heat incentive will soon be implemented in the UK for the first time and is planned in the Netherlands.

(5) Developing properties certified with a 'green' building label

Independent of policy incentives, a business case also exists if a property developer can achieve a higher sales price for a building which is certified according to a voluntary 'green' building label. This is frequently the case in the North American and some Asian markets.

(6) Building owner profiting from rent increases after the implementation of energy efficiency measures

For building owners who do not occupy the building themselves and for housing corporations, revenue opportunities from an investment in energy efficiency arise when they are allowed to charge a higher rent from the tenants after the renovation. The higher rent takes the tenant's energy savings into account. The required changes in the legal framework address the issue of split incentives.

Business models based on new financing schemes

High up-front costs are a major barrier for an increased deployment of RET. Innovative financing schemes may therefore create business cases, if the financing schemes help to overcome the barrier of high up-front costs. As public budgets are limited, new and innovative financing schemes are emerging which do not burden government budgets.

(7) Property Assessed Clean Energy (PACE) financing

The Property Assessed Clean Energy (PACE) concept has for example been widely discussed and piloted in the US. Under this scheme, local governments issue bonds for RET projects. The building owner repays the loan through an additional special assessment payment on its property tax bill for a specified term (Institute for Building Efficiency, 2010b). When the property changes ownership, the remaining debt is transferred with the property to the new owner.

(8) On-bill financing

On-bill financing programmes are another model for addressing the barrier of high up-front costs and access to capital. A utility provides capital to a home owner for the installation of RET or EE measures. The home owner repays the investment via its energy bill.

(9) Leasing of RET equipment

Leasing of RET offers another opportunity for building owners to use RET without having to make an up-front investment. It is possible both for larger-scale equipment in large commercial buildings and in some cases also for small-scale, innovative RET for private home owners. The opportunity to lease equipment may also be part of the energy services package offered by an ESCO. However, leasing of RET equipment is analysed separately as it is technology specific and may also target individual residential customers.

(10) Business models based on energy saving obligations

Innovative financing options can also emerge under energy saving obligations for utilities. The utility (potentially via an ESCO) offers investment incentives for energy efficiency investments, which are financed by overall higher energy prices. These incentives offer opportunities for building owners.

Chapter 4

Analysis of business models

The following describes and analyses the ten business models in more detail. The analyses of the business models based on new revenue models and on new financing scheme follow the same template (including an introduction and definition, applicable technologies, market segments, involved actors, organisational and financial structure, existing policy and market context, analysis of strengths, weaknesses, opportunities and threats (SWOT), discussion and conclusions). To avoid lengthy repetitions of information, for the energy contracting (EC)/ESCO business models in the category product service systems, first common features of all EC business models are described, followed by a description of the three individual ESCO models. These descriptions are wrapped up with a SWOT analysis, and discussion and conclusions for the EC models in general.

Product service systems: energy contracting (ESCO or energy efficiency services)

Introduction, definition and common key features of all three ESCO models

Introduction and definition

Energy contracting (EC) – also labelled as energy service companies (ESCOs) – is one of the most prominent examples of product service system business models for sustainable energy. Two basic ESCO business models can be distinguished, which provide either useful energy (energy supply contracting – ESC) or energy savings (energy performance contracting – EPC) to the end-user. In addition to the two basic models, a hybrid model labelled as integrated energy contracting (IEC) was introduced and is being piloted in Austria and Germany. IEC aims to combine useful energy supply, preferably from renewable sources with energy conservations measures in the entire building.

Independent of the business model, energy services – in a more narrow sense – have several common features, which are outlined in this first section.

Most importantly an ESCO's remuneration is performance-based (it is paid for the measured outputs as opposed to the inputs consumed) and it guarantees for the outcome and all-inclusive cost of the service package. All ESCO business models investigated here lead to a reduction of final energy demand. In addition, they achieve environmental benefits due to the associated energy and emission savings in addition to non-energetic benefits such as an increase in comfort or reputation gains.

Various definitions of energy services can be found in respective standards and literature.[1] However most commonly applied definitions fall short with regard to important properties of 'real' energy contracting services. Such properties are outsourcing of commercial and technical risks to an ESCO, guarantees for results and 'all-inclusive' costs of the measures implemented or of the optimisation according to project cycle costs. These features may constitute an added value in comparison to standard in-house implementation of energy services. Therefore, in a narrow sense we define energy contracting (EC) as:

> Energy contracting is a comprehensive energy service concept to execute energy efficiency projects according to minimised project cycle cost.
>
> Typically an energy service company (ESCO) acts as a general contractor and implements a customised service package (consisting of e.g. design, installation, (co-)financing, operation and maintenance, optimisation, fuel purchase, user motivation).
>
> As key features, the ESCO's remuneration is performance-based, it guarantees for the outcome and all-inclusive costs of the services and takes over commercial as well as technical and operational risks over the project term.
>
> (after Bleyl & Schinnerl, 2008a)

In addition to 'real' energy contracting models, there are so-called energy service providers which offer technical and engineering services for the identification and implementation of RET and EE projects, but do not offer any performance guarantees (see e.g. examples in Box 5.1). Figure 4.1 shows a value chain of different types of energy service activities.

Modular scope of services

EC services are not about a particular technology or energy carrier. Instead EC is a flexible and modular 'tool' to execute energy efficiency and RET projects according to the goals of the facility owner.

All the tasks shown in Figure 4.2 such as planning, construction and financing, the ongoing components of the service package (operation and maintenance, purchasing of fuel, quality assurance and measurement and verification) as well as compliance with the legislative framework have to be

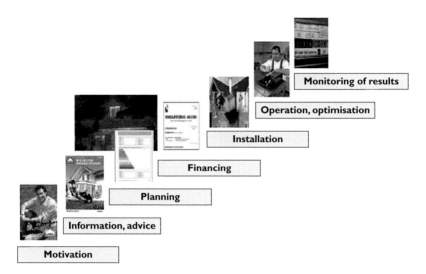

Figure 4.1 Potential energy service activities along the value chain of energy efficiency improvements in buildings (Source: Leutgöb et al., 2011)

covered either by the building owner or the ESCO throughout the contractual period.

For implementation, the building owner assigns a customised energy service package and demands guarantees for the results of the measures taken by the ESCO. The necessary components for implementing energy projects are summarised in Figure 4.2.

Typically an ESCO serves as a general contractor and is responsible for coordination and management of the individual components and interfaces

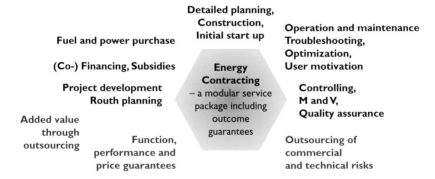

Figure 4.2 Energy contracting: a modular energy service package with guaranteed results for the client. (Note: The added value for the client of energy contracting compared to in-house implementation is displayed in red.)

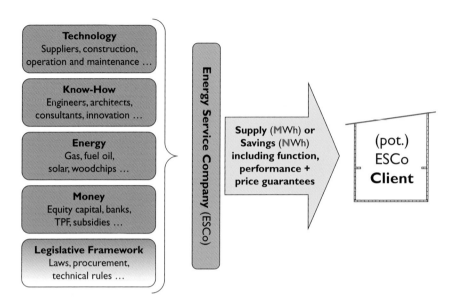

Figure 4.3 Energy contracting: components of service package and outsourcing of interfaces and guarantees to an ESCO

of the service package towards the customer. It has to deliver the commissioned energy service (Megawatt hours of useful energy or energy savings ('Negawatt hours')) to the customer at 'all-inclusive' prices as displayed in Figure 4.3.

Energy efficiency projects differ in their contents and general conditions. Therefore, it has proven to be necessary and sensible to adapt the scope of services specifically to the individual project. This also implies that the building owner can define which components of the energy service are outsourced and what he carries out himself (e.g. ongoing on-site maintenance provided by a facility manager or financing from other sources).

An important difference between in-house ('do-it-yourself') implementation and outsourcing to an ESCO is the functional, performance and price guarantees provided by the ESCO and the assumption of technical and economic risks by the ESCO.

Actors

Directly involved actors are the ESCO and the building owner. The second layer in the value chain includes equipment and final energy suppliers and financial institutions, who provide capital for the investment into (RET and EE) equipment. No direct policy intervention is required.

Financing of the required investment

Outsourcing of up-front financing of RE or EE equipment is often the key driver to engage with an ESCO. However ESCOs are not necessarily able to offer more attractive financing conditions in comparison to a building owner, especially when the client is a large organisation, nor is financing typically the ESCO's core competence. Therefore, the ESCO service package does not necessarily need to include financing. Financing can be provided by the building owner (Figure 4.4), the ESCO (Figure 4.5) or a third financing partner, depending on who has better access to capital and financing conditions.

This distinction is also reflected in the Anglo-Saxon EPC markets, where two basic EPC models are differentiated mainly with regard to who finances the investment: 'Guaranteed Savings' refers to a service model without ESCO finance, whereas the 'Shared Savings' model includes financing in the ESCO's service package.

Combinations of the above options are also possible to account for the specific project and the actors involved. In reality, a mixture of financing sources is often the best choice in order to balance risks. If the ESCO does not provide financing itself, it can still take on the role as a facilitator supporting the building owner to get access to third party financing solutions.[2]

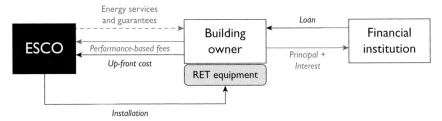

Figure 4.4 Energy contracting model where building owner finances RET equipment through a loan from a financial institute

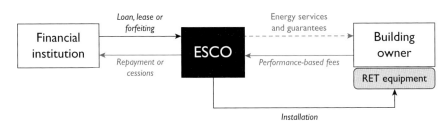

Figure 4.5 Energy contracting model where ESCO finances RET equipment (optionally with financial institute)

Existing markets and policy context

Reliable market data on ESCO markets are scarce or not publicly available. In the EU, energy supply contracting (ESC) has by far the largest market share within the energy services sector (Labanca, 2010).

In Germany, ESC has for example a market share of 85–90 per cent of the ESCO market (Bleyl, 2011). While energy performance contracting is talked about a lot, its market share in the German ESCO market is only between 10 and 15 per cent (Prognos, 2009; VfW, 2009). In Germany, the most recent market estimate indicates that there are about 250 companies active in the energy services market, mostly using the ESC model (Eikmeier et al., 2009). The total volume of the German energy services market is estimated to be about €2 billion annually, of which about 60 per cent takes place in the residential buildings (Bunse et al., 2010).

Integrated energy contracting (IEC) is an innovative model which has been piloted in Austria. Experiences collected from, up to now, eight projects have confirmed the practical feasibility of the IEC model. Beyond that, it remains to be seen what contribution IEC will make as a tool for the implementation of sustainable energy projects (Bleyl, 2011).

In order to support energy services in general, in the EU, a large number of countries propose or have implemented policies and supporting measures, such as information campaigns (Boonekamp & Vethman, 2010). One of the drivers for such legislation is the EU Directive on Energy End Use and Energy Services (2006/32/EC). There seem to be no specific policies directly supporting specific energy contracting models, such as ESC or EPC (Szomolanyiova & Sochor, 2012). In Japan, the 2007 New Procurement Law for the Environment encourages authorities to procure ESCO services for public buildings (WBCSD, 2008). It is expected that the policy context of a country does play an important role in the development of the energy services market. In Denmark for example, energy services are implemented because energy saving obligations for energy suppliers may only be implemented by third party energy service companies.

For the market development of EPC, a key enabling factor has been the involvement of independent third party organisations, acting as market and project facilitators between potential customers and ESCOs. Often they are energy agencies (e.g. Grazer Energie Agentur[3] or Berliner Energieagentur) or consultants,[4] who develop concrete EPC or ESC projects, mostly on behalf of the client, prepare calls for proposals and model contracts and put them out on the market for bidding.

In some parts of the world, so called public 'Super ESCOs' have been proposed or implemented, e.g. Energy Efficiency Services Limited in India, FEDESCO in Belgium or HEP ESCO in Croatia. The scope of their (planned) activities is extremely broad and ranges from market and project facilitation for (potential) clients and ESCOs to acting as a full-fledged ESCO themselves. Moreover, these organisations may be tasked amongst others to solve financing bottlenecks and undertake general information campaigns (Limaye, 2011). The success of this broad concept remains to be seen. The portfolio of Super ESCOs may require a more focused approach, particularly regarding market development and project facilitation activities.

Energy supply contracting

Description

Under an energy supply contracting (ESC) model, an energy service company supplies useful energy, such as electricity, hot water or steam to a building owner or building user (as opposed to final energy such as pellets or natural gas in a standard utility contract). The output is measured and verified in Megawatt hours delivered. ESC models run under long-term contracts of typically 10 to 15 years, depending on the technical lifetime of the equipment deployed.

Extended project terms or building cost allowances allow including measures with longer payback times like facades with integrated PV modules or entire building shells as well.

This business model gives the building owner the opportunity to outsource technical and economical risks associated with energy supply related activities, including the planning, installation, operation and maintenance and financing of equipment for heating, cooling or electricity generation to a professional party and to buy services instead of individual components. ESC often includes supply of final energy through the ESCO, however on its own accounts. The standard scope of services is limited to the energy supply side, e.g. the boiler room in the basement of a building but may very well include solar supply options or other RET as displayed in Figure 4.6.

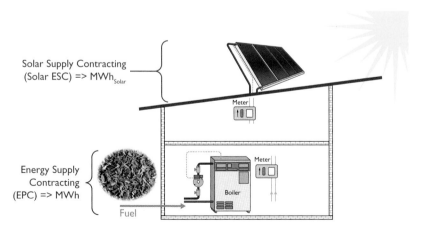

Solar Supply Contracting (Solar ESC) => MWh$_{Solar}$

Meter

Energy Supply Contracting (EPC) => MWh

Meter

Boiler

Fuel

Figure 4.6 ESC-model: schematic standard scope of services including renewables

The ESCO's remuneration is performance-based and depends on the useful energy output delivered. Thus the ESC model provides an incentive to increase efficiency of the final energy conversion and to reduce primary energy demand. It guarantees for the outcome and all costs of the services and takes on the commercial as well as technical and operational risks of the project. ESC may accelerate the uptake of RET, if RET are cost-competitive over the life-cycle of the project because ESCOs have an inherent interest to reduce life-cycle costs.

Market segments

ESC is applied in different end-use sectors such as housing, commerce, industry or public buildings. For the housing sector specifications of minimum project sizes to be economically viable exist: Eikmeier et al. (2009) detail a thermal load of 100 kW as a lower threshold based on a transaction cost analysis and empirical results from a market query. In a simple approximation this corresponds to annual energy costs of about €20,000. Upper project sizes are not limited and may go up to 10 MW or more for large industrial installations and encompass supply of heat, steam, (back-up) electricity or compressed air.

Applicable technologies

ESC supplies useful energy such as hot water, steam, (back-up) electricity or compressed air from a wide variety of technologies based on conventional or renewable sources.

Technologies applied typically are efficient boilers ((bio-)gas, wood chips and pellets), combined heat and power (CHP) systems (gas turbines and reciprocating engines), district and small-scale heating networks, solar thermal and solar PV installations. ESC is particularly suitable for the implementation of RET like solar or geothermal applications, as their energy outputs can usually be measured with little effort through electricity or heat meters.

Organisational and financial structure

The ESCO is responsible for the implementation and operation of the energy supply package at its own expenses and risk, according to the project specific requirements set by the client. In return, the ESCO is remunerated for the useful energy delivered, depending on the actual consumption in combination with a flat rate for operation and maintenance. The business model is displayed in Figure 4.7.

The ESCO's remuneration is made up of the following three price components (see Figure 4.7):

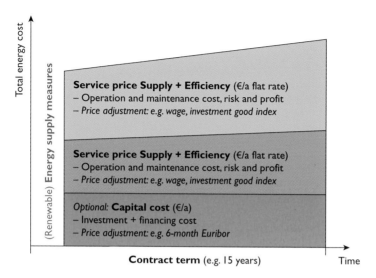

Figure 4.7 Energy supply contracting business model

1   Energy price (per MWh of useful energy metered), which covers the marginal 'consumption related' cost per MWh of useful energy supplied. To account for final energy price developments during the contractual period, the ESCO's energy price will be adjusted by using statistical energy price indices depending on the fuel used (e.g. gas or biomass index). Thus, the risk related to final energy price development remains with the ESCO's client. In order to rule out incentives to sell more energy, the ESCO's calculation of the energy price should include consumption related cost only (the marginal costs), i.e. exclusively the expenditure for fuel and auxiliary electricity. If the energy price is at the marginal cost, there is no incentive for the ESCO to sell more, because the price equals the ESCO's costs.

2   The service (or basic) price for energy supply (flat rate) includes all operational cost, i.e. the cost for operation and maintenance, personal, insurance, management etc. of the energy supply infrastructure as well as entrepreneurial risk.

    During the contractual period, the prices are usually adjusted (typically every year retrospectively) by using statistical indices such as wage or investment good indices. The service price for energy efficiency (flat rate) is determined in analogy to the above service price including all operational cost of the energy efficiency measures. As shown in Figure 4.7, the two basic prices can be combined.

3   If the ESCO (co)-finances the equipment its remuneration also includes a fee for its capital costs minus any subsidies for the RET equipment which it may have received.

In the above mentioned price components, all of the ESCO's expenditure items for the defined scope of services throughout the contractual period must be included ('all-inclusive prices').

Discussion and conclusions

The ESC model is a proven model to implement efficient supply from fossil and renewable sources in new and existing public, industrial, commercial and large residential sector buildings. It is effective in reducing final energy demand, because the ESCO pays for the final energy needed and is remunerated for its useful energy output only. However, efficiency gains are usually limited to the energy supply system.

The following are the main conclusions for the ESC model:

- ESC is particularly suitable for the implementation of RET like solar or geothermal applications, because their energy outputs can usually be measured with little effort through electricity or heat meters. In comparison to the EPC model, ESC reduces the expenses for measurement and verification and the risks associated with the savings guarantee significantly.
- However, large demand side energy efficiency potentials remain untapped, because the scope of services is limited to the provision of useful energy.
- Transaction costs for ESC projects require a minimum project size, which can be expressed as a minimum energy cost baseline of about €20,000. The ESC model is thus not suitable for individual or small multi-family houses.
- In order to rule out incentives to sell more energy, the energy price component should be set at marginal cost. This implies that the ESCO's calculation of the energy price should include variable cost only, i.e. exclusively the expenditure for fuel and auxiliary electricity. If the energy price is at marginal cost, there is no incentive for the ESCO to sell more energy, because the price equals its cost.
- No energy cost baseline is needed for the business model to work. If desired by the building owner, savings achieved can still be calculated by comparing to a historic or calculated baseline.

Energy performance contracting

Description

Under an energy performance contracting (EPC) business model, an energy service company guarantees energy cost savings (also labelled as 'Negawatt-hours') in comparison to a historical (or calculated) energy cost baseline. For its services and the savings guarantee the ESCO receives performance-based remuneration in relation to the savings it achieves.

Generally, savings achieved can only be measured indirectly as difference between consumption before and after implementation of the EE and RE measures (relative measurement: savings = baseline – ex post-consumption) (for more details please refer to the 'organisational and financial structure' section).

The standard scope of services encompasses the entire building as displayed in Figure 4.8. RET may play a role but with most EPC projects the main focus is on the implementation of energy conservation measures.

EPC models run under long-term contracts of typically ten years, depending on the payback time of the energy savings measures and the specification of the building owner.

Market segments

The market for EPC projects is currently largely limited to public institutions at federal, state and regional levels including special purpose buildings like universities, hospitals and leisure facilities. In Germany, for example, projects are spread very unevenly. EPC projects are found particularly in cities or regions where independent market and project facilitators such as energy agencies engage on behalf of buildings owners in preparing concrete projects and putting them out on the market for ESCOs to bid for. One such example is the Berlin Energy Saving Partnership described in Appendix A.12. US market data show a similar picture: 84 per cent of ESCOs' revenues from EPC projects stem from public institutions, consisting of federal buildings and so called 'MUSH' markets (municipal and state governments, universities and colleges, K-12 schools and hospitals) (Satchwell et al., 2010).

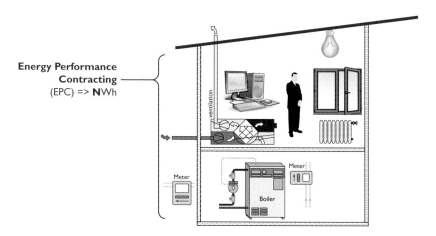

Figure 4.8 EPC model: schematic standard scope of services

Transaction and measurement and verification costs of EPC projects are high. As a consequence the EPC market is dominated by large projects. A minimum energy cost baselines can be set at 100,000 €/a, but realised projects are typically an order of magnitude above. The 24 pools of buildings of the Berlin 'Energy Saving Partnership' – one of the most successful EPC campaigns in Europe – have for example an average energy cost baseline of €1.88 million/year (ESP Berlin, 2009; see also Appendix A.12).

Applicable technologies

An EPC contract may feature savings for all energy carriers such as electricity, gas or water. Typical measures are energy management and controls, heating, ventilation and air conditioning (HVAC)-technologies like air conditioning systems, hydraulic adjustment of distribution networks or lighting. Sometimes an exchange of boilers or adjustment of district heating connections is also undertaken. In addition, the scope of services frequently also includes influencing the behaviour of building occupants through information campaigns and incentive programmes.

Indications for the potential for energy efficiency improvements that may be unlocked through EPC contracts can be derived from realised, large-scale EPC projects: the 'Energiesparpartnerschaft' in Berlin and the 'Federal Contracting Campaign' in Austria, for example, both report savings between 20 and 25 per cent (ESP Berlin, 2009; BMWFJ, 2012).

Organisational and financial structure

The ESCO is responsible for the implementation and operation of the energy efficiency package at its own expense and risk, according to the project specific requirements defined by the client and the ESCO. Purchasing of final energy (electricity, fuels) mostly remains with the building owner. The standard business model scheme is displayed in Figure 4.9.

The ESCO's remuneration in an EPC model is often labelled as 'contracting rate'. It is usually calculated as a percentage of the savings achieved through the EE and RE measures. In case of under achievement the ESCO needs to compensate for the losses, but it will receive an additional remuneration in case of over achieving the savings guarantee.

After the end of the contract term, the facility owner benefits from the full energy cost savings, but all operation and maintenance expenses are on his accounts.

The contracting rate needs to cover all expenses of the ESCO for the defined scope of services throughout the contractual period ('all-inclusive prices'). Typically this includes the implementation of the measures, their operation and maintenance, pre-financing of the investment and taking over risks according to the project specifications defined in the contract. If the ESCO (co)-finances the equipment, the remuneration must also cover capital costs.

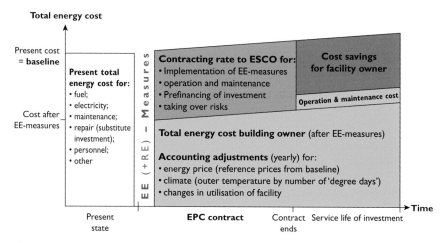

Figure 4.9 Energy performance contracting business model

When measuring savings through a comparison between a baseline and post-retrofit energy costs, two major difficulties may occur:

- The baseline itself may be difficult to determine with enough accuracy due to a lack of availability of historic data (e.g. from bills or meters).
- The determined energy cost baseline is not a constant but subject to changes in climate conditions (e.g. ambient temperatures, solar radiation etc.) and in energy prices. Besides, utilisation of the building may change. These changes need to be taken into account when calculating energy cost savings; especially the changes in utilisation may cause considerable difficulties for the ESCO and the facility owner in adjusting the baseline.

In addition to the resources necessary (high transaction and operational costs), the baseline determination and adjustment can cause a considerable degree of insecurity and monetary risks for the (prospective) project partners. Determining and adjusting the baseline is a crucial issue in the EPC business model and needs to be undertaken for all performance-based billing periods over the entire contract term. The aforementioned difficulties and risks underline the necessity for a clearly defined measurement and verification plan for each EPC project (see e.g. IPMVP, 2009).

Discussion and conclusions

The following are the main conclusions for the EPC ESCO model:

- EPC provides a comprehensive approach to end-use efficiency improvements. RET may play a small role.

- With a market share of about 10 per cent of the ESCO market, the market uptake of EPC is significantly lower than for ESC. The market is mainly limited to the public sector and special purpose buildings such as hospitals, swimming facilities or universities.
- Today the EPC model is applied for large projects only, with minimum energy cost baselines of €100,000 per year and markedly above, among other reasons because transaction as well as measurement and verification costs of EPC projects are high.
- Determining, measuring and verifying a baseline and the appraisal of risks and costs of the savings guarantee hinder a more widespread market uptake. There is a widespread expectation that EPC projects must be completely re-financed from future energy cost savings only and in addition create immediate cost savings. This is achievable only for projects with very high savings potentials and short payback periods, thus severely limiting the application of EPC as an energy efficiency tool.
- The initiation of policy supported implementation programmes such as the Federal Energy Management Program (FEMP) in the US, the Berlin Energy Saving Partnership (see Appendix A.12) or the Federal Contracting Campaign (BundESCOntracting) in Austria are an important enabling factor for the growth of EPC projects. As a consequence, higher market penetrations are particularly observed where independent market facilitators such as energy agencies engage on behalf of building owners in preparing concrete projects and putting them on the market for ESCOs to bid for.

Integrated energy contracting

Description

The integrated energy contracting business model is a hybrid of ESC and EPC and combines two objectives:

1 Reduction of energy demand through the implementation of energy efficiency measures in the areas of building technology (HVAC, lighting), building shell and user behaviour.
2 Efficient supply of the remaining useful energy demand, preferably from renewable energy sources.

IEC is based upon the widespread energy supply contracting business model and is supplemented by quality assurance instruments and deemed savings approaches for the energy efficiency measures.[5] The latter serves as a substitute for the potentially complex and costly measurement and verification of energy savings undertaken in the EPC business model. Therefore IEC reduces transaction costs particularly for smaller projects.

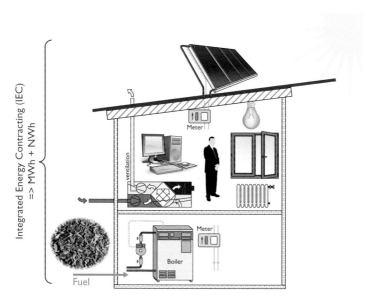

Figure 4.10 IEC model: schematic standard scope of services

As compared to standard energy supply contracting, the range of services and thus the saving potential to be utilised is extended to the overall building or commercial enterprise (see Figure 4.10). The scope is not limited to the supply of heat energy. Instead, the model is intended to be used for all energy carriers and consumption media such as heat, electricity, water or compressed air.

As with ESC and EPC, the IEC business model offers the building owner the choice to outsource technical and economical risks associated with the implementation of RET and EE measures to a professional third party and to buy services instead of individual components. IEC may turn out to be particularly suitable to combine supply from renewable sources with energy conservation measures and thus accelerate the uptake of RET, provided RET are cost-competitive over the life-cycle of the project because ESCOs have an inherent interest to reduce life-cycle costs.

Market segments

Since IEC builds on the ESC model, a similarly wide range of end-use sectors such as commercial and public buildings as well as the residential sector can be targeted. For more details please refer to the ESC model description.

Applicable technologies

IEC combines energy efficiency and RET measures. All technologies listed in the ESC and EPC business model descriptions are applicable.

Organisational and financial structure

The ESCO is responsible for the implementation and operation of the energy efficiency package at its own expense and risk, according to the project specific requirements defined by the client and the ESCO. Purchasing of final energy (electricity, fuels) mostly remains with the building owner. The standard business model scheme is displayed in Figure 4.11.

Basically the IEC business model builds on the ESC with similar price components and is supplemented with a flat rate price for the energy efficiency measures. To avoid or at least to reduce the (potential) EPC problems, the supposedly exact measurement and verification of the actual savings under an EPC approach is replaced by quality assurance and simplified measurement and verification procedures (e.g. deemed savings).

The individual quality assurance instruments (QAIs) for the installed EE measures secure the functionality and performance of the measures, but not their exact quantitative outcome over the entire project cycle. The objective is to simplify the business model and to reduce (transaction) cost by balancing measurement and verification cost and accuracy. Appropriate QAIs need to be defined for each EE measure, e.g. a one-time performance measurement for a new street lighting or a one-time thermographic analysis for verifying the quality of a refurbished building shell. These QAIs replace the annual measurement and verification of the EPC savings guarantee.

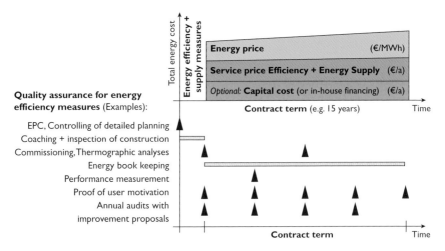

Figure 4.11 Integrated energy contracting business model

Discussion and conclusions

The IEC business model builds on the energy supply contracting (ESC) model, which is known and applied in public, residential, commercial and industrial buildings. The scope of services and thus energy savings potential is extended to the overall building or enterprise and to all consumption media, such as heat, electricity and water. At the same time methodological problems of energy performance contracting (EPC) as mentioned earlier, e.g. related to creating and adapting baselines, are avoided or at least reduced, e.g. by avoiding the need for a baseline and its adaption in the course of the project.

The following are the main conclusions for the IEC ESCO model:

- IEC allows for combining energy savings and supply of energy in an integrated approach. Therefore, in comparison to standard ESC, higher end-use energy saving potentials can be achieved. Moreover, RET may play a key role.
- IEC is an innovative model. Practical experiences are still limited, but eight pilot projects in Austria have proven the feasibility of the model (see case study LIG, Austria in Appendix A.1) Some experienced ESCOs have expressed interest in developing own products based on the IEC model.

For additional conclusions, please refer to the ESC model.

SWOT analysis, discussion and conclusions for all three ESCO models

SWOT analysis

This analysis summarises important findings for the deployment of RET and EE measures. If not mentioned otherwise, the analysis mainly takes the building owner's perspective. Most implications concern ESCO business models in general, otherwise they are marked with the respective EC-model acronyms.

STRENGTHS

For the building owner:

- ESCOs will invest into RET and EE measures, if they are cost-competitive over the contract term, because the ESCO has an intrinsic incentive to reduce life-cycle costs.
- Building owners pay for outputs and results (services) instead of inputs and components (e.g. technology). Thus technical as well as financial and operational risks can be outsourced to an ESCO and the building

owner can request guarantees for the total cost and overall performance of the energy service package.

- Energy contracting models can facilitate access to capital to overcome high up-front cost of RE and EE investments. Some ESCOs provide financing themselves, but frequently ESCOs are capital constrained but may still take the role of facilitator for third party financing solutions.
- EC is a modular and customised service package according to the specifications of the building owner.
- Outsourcing the responsibility for energy related services to an experienced actor may reduce information barriers, up-front cost and access to capital (if the ESCO (co)-finances the equipment or facilitates financing) and the 'hassle factor' for the building owner.
- *ESC and IEC* are particularly suitable for RET, because their energy outputs can be measured directly without needing a baseline. Thus, in comparison to the EPC model, the expenses for measurement and verification and the risks associated with the savings guarantee are significantly reduced.

General:

- EC projects are market-based with relatively little dependence on supporting policy measures. RET and EE measures in one building as well as several buildings can be packaged in order to reduce transaction cost. In this package or building pool, 'low hanging fruits' may support higher hanging ones like RET to make the whole package economically feasible.
- Optimisation of life-cycle costs in an energy contracting model may lead to deployment of RET and EE measures which may not be undertaken by the building owner alone. ESCOs may lift several barriers like separate budgets for investment and operation, lack of financial and human resources, and knowledge of available incentive schemes.

WEAKNESSES

- Energy contracting is limited to cost-effective measures. But subsidy schemes can be integrated in the models either through the building owner or the ESCO.
- If the ESCO is responsible for investing in equipment with long pay-back times, long contracting periods are required which result in mutual long-term dependencies and require a long-term business perspective of the ESCO and the client. For premature contract termination, buy-out clauses can be agreed in the contract.
- ESCO projects require minimum project sizes, which can be expressed in minimum annual energy cost baselines. Today the *EPC* model is applied for large projects only with minimum energy cost baselines of

€100,000 per year and markedly above. For *ESC* minimum energy costs are about 20,000 €/year.

- The scope of services of an ESC scheme is limited to the energy supply. ESC does not maximise the full potential for energy efficiency improvements and CO_2 reductions in the building.
- EC-models are complex. They cover the entire project life-cycle in one contract and require technical, economical, financial, legal and organisational know-how. In this context the role of independent market and project facilitators has proven to be key to overcome the challenges related to the complexity, particularly for EPC projects.

OPPORTUNITIES

- *ESC* and *IEC* are suitable for smaller projects (in comparison to EPC), and thus have a larger market potential. The *ESC* model is known and applied in residential and commercial housing and industry in addition to public buildings.
- In many countries around the world, ESCO markets are growing as building owners realise the added value of outsourcing activities related to sustainable energy.
- Supportive policy measures, for example energy saving obligations for energy suppliers, are expected to lead to a growth of the market for energy services.
- With rising fossil fuel prices, RET and EE measures deployed by ESCOs become more attractive.

THREATS

- ESCO business models depend on the willingness of a building owner to outsource comprehensive service packages. Outsourcing may threaten existing jobs, organisational routines and even question the performance of individuals previously responsible for sustainable energy agendas. Consequently ESCO models may face opposition from existing personnel of building owners, because changes in competences, organisational and procurement routines are required.
- Hiring sufficiently qualified personal with interdisciplinary skills may be a barrier for ESCO development.
- Although EC models are a market-based instrument, some (legislative) policy support is required to solve existing barriers, e.g. by:

    - Allowing public entities to conclude multi-year contracts with ESCOs, which do not count against public deficit limits.
    - Addressing the barrier of 'split incentives' between building owners

and renters/occupants. This applies particularly to the residential but to a lesser extent also to the commercial building sector.

- Allowing life-cycle cost optimisation across separate investment and operational budgets. This is a key barrier for private and public organisations.

Discussion and conclusions for all ESCO models

The EC service concept shifts the focus of energy supply and management from buying or selling units of final energy (like fuel oil, gas or electricity) towards the desired benefits and services derived from the use of the energy carrier (e.g. keeping a room warm, air-conditioned or lit).[6] EC is an instrument to minimise life or project cycle cost, which takes the operation of the building into account. As the ESCO's remuneration depends on the output of the services provided and not the inputs (like fuel or man-hours) consumed,

Strengths
- Proven and market based model
- Performance-based payments provide incentives to maximise efficiency
- Reduces 'hassle factor' for building owner by outsourcing risks, guarantees for all-inclusive costs, and modular service package

Weaknesses
- Limited to cost-effective measures
- Long contracting periods and minimum project sizes required
- ESC is limited to energy supply, does not maximise the full EE potential in a building

Opportunities
- Expected growth of ESCo markets with increased awareness of the benefits of energy contracting
- Increasing cost-competitiveness of RET,
- Regulatory support and increased engagement by public building owners

Threats
- Willingness of building owners and existing personnel to outsource service package to an ESCo
- Complex contracts covering entire project cylce
- Separate investment and operational budgets of building owners
- Split incentives

Figure 4.12 Energy contracting models: summary of SWOT analysis

the concept induces an intrinsic interest for the ESCO to increase efficiency and to reduce final energy demand.

For building owners, energy contracting models offer the opportunity to outsource activities related to (sustainable) energy, including the planning, installation, operation and maintenance, and financing of equipment for heating, cooling or electricity generation to a professional third party. EC reduces the need for internal capacity to deal with these issues and for managing a variety of different suppliers and interfaces, thus allowing building owners to concentrate on their core business. However, building owners still need a certain understanding of energy related issues in order to negotiate contract conditions and supervise the ESCO.

Figure 4.13 summarises the two basic ESCO business models along the value chain from primary energy to energy services. The figure also shows standard products and services offered under the different business models.

ESCO models are well replicable, but they do require specialist knowledge to operate the business model. In general, the ESCO implementing the scheme will have a much better knowledge of available RET, their characteristics, relevant suppliers, and available incentive and financing schemes than a building owner. Therefore one of the major advantages is that the ESCOs is more likely to take advantage of opportunities for the deployment of RET, leading to an increase in deployment. Such opportunities will increase with increasing competitiveness of RET versus conventional technologies and if additional incentive schemes are implemented.

The following main conclusions are applicable to all energy contracting models covered in this book:

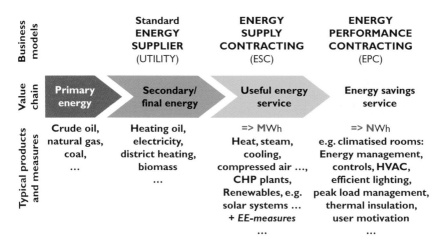

Figure 4.13 Value chain from primary energy to energy services[7]

- ESCO models have an intrinsic incentive to minimise life respectively project cycle cost over the entire contract term of typically 10 to 15 years. In the ESCO's price, all expenditure items for the defined scope of services throughout the contractual period must be included ('all-inclusive prices').
- When comparing between outsourcing to an ESCO versus in-house implementation, the functional performance and price guarantees provided by the ESCO and the assumption of technical and economic risks through the ESCO may constitute an added value for the client and should be taken into account in addition to the comparison of total costs of different options.
- For ESCO models, different options for financing the RE and EE equipment are feasible. The financing structure should be chosen depending on opportunities for access to capital and risk preferences of the involved actors and the specifications of the project. Often, a mixture of financing sources is the best choice to balance risks, where ESCOs are used as a vehicle and facilitator for third party financing, but not necessarily as financiers themselves.
- ESCO business models cannot substitute the client's basic decision to implement RET and EE measures. This decision remains a prerequisite for the success of ESCOs.
- The decision to outsource energy services can only be taken by building owners. For the public housing stock, however, governments have the opportunity to directly support energy contracting business models. This can be done by changing public procurement rules to allow or require decision makers in public buildings, including housing corporations, to procure equipment according to lowest life or project cycle cost (net present value), and by allowing them to enter into long-term contracts with an ESCO.
- Compared to other energy services models, in particular energy performance contracting (EPC), ESC is particularly well suited for generating electricity and heat from RET, as the output of Megawatt hours delivered can be measured relatively easy, thus reducing measurement and verification cost in comparison to the EPC model.
- Long-term experience with different ESCO markets shows that the development of comprehensive energy (RET and EE) projects is strongly supported by facilitators with a sufficient time horizon to raise awareness and commitment with stakeholders. Independent of the ESCO business model, the decision (voluntary or driven by regulation) of the building owner to invest in RET/EE measures remains a basic requirement for the involvement of an ESCO. Furthermore, ESCOs are not able to address the barrier of the fragmented nature of the building sector which leads to a large number of small units of energy saving potentials. However, there may be significant potential for companies, e.g. individual installers

or groups of them, in offering less extensive energy services to individual households.

Business models based on new revenue models

Making use of a feed-in remuneration scheme

Description

A feed-in scheme is a policy by which the producer of renewable energy receives a direct payment per unit of energy produced. This feed-in remuneration can be a *tariff*, which like a preferential price covers the full generating costs, or it can be a *premium*, which provides a 'bonus' for the producer to cover the financial gap between the generation costs of using renewable energy versus using conventional (fossil) energy. A feed-in scheme guarantees access to a predictable and long-term revenue stream, which can serve as a stable basis for a business model.

Feed-in schemes have been used by companies and investors as a basis for business models for large-scale power production (e.g. wind parks or biomass plants), as well as by households and small and medium enterprises (SMEs) who want to generate their own energy using renewable sources (e.g. solar PV or biomass heating). Such business models by households or SMEs can focus on production for own use, or for the sale of energy to the grid (see case study 'Greenchoice' in Appendix A.5) (or for heat to a nearby user). Feed-in schemes typically differentiate in categories by size of the installation, technology and fuel used. The level of remuneration is based on the category specific generation costs, but the actual payment is based on production (Gifford et al., 2011).

Market segments

Feed-in based business models are applicable for all market segments: new and existing buildings, public, commercial/industrial and residential buildings. Notably, in the domestic building segment, a feed-in scheme may provide opportunities for entrepreneurs who use demand aggregation (e.g. through district heating or by providing energy services to groups of customers). Which market segments are eligible for feed-in support, and therefore can be part of a business model depends on the policy specifics in the country or region.

Applicable technologies

In principle, feed-in schemes, and associated business models, can be designed for all technologies that generate heat or power (or both). In practice, the tariff or premium is based on the estimated generation costs over the lifetime of the installation, which makes it most suitable for technologies that are available off-the-shelf, e.g. solar PV for electricity generation or heat pumps for heat production, and less so for innovative (and diverse) or unique technologies. Some schemes primarily cover electricity production and an additional 'bonus' is made available if associated heat is put to productive use.

Actors

The two main actors in a feed-in scheme are the institution that makes the payment available (government, network operator) and the recipient (home owner, building manager, or energy service company). The actual payment can be executed through a government agency, the energy supplier, or through the network operator. Payment in many instances is based on certificates, or 'guarantees of origin', in which case a government agency or certified third party company will be involved in verifying production and issuing certificates. If tariff levels are based on category specific generation costs, an (independent) institute may be involved in advising the government on costs.

Organisational and financial structure

A feed-in scheme is a policy, and the tariffs (and budgets) are therefore fixed by the government. The cost of this support is either recovered from the government budget (i.e. from tax payers), or from a network operator mark-up on energy bill (i.e. from energy consumers, as is the case for the German scheme).

In addition to the *tariff scheme*, in which the producer gets a fixed price for the supply of energy, and the *premium scheme*, in which the producer gets a premium in addition to income from selling the energy on the market, hybrid forms are also possible. In these, the premium is chosen to complement the income from the market, to jointly cover the generation costs. So, in practice, the producer gets a fixed amount, but from two different sources. This hybrid form is similar to a feed-in tariff for the investor, but it has different consequences for government expenses.

A feed-in scheme typically publishes rates per energy unit for eligible production (e.g. in €/GJ or €/kWh). If a producer is eligible, a contract (or agreement) can be obtained from the government which allows the producer to claim the specific tariff (or premium) for every unit produced. This agreement fixes the conditions and levels of the tariff, typically 8–20 years depending on the technology. So once an agreement is entered into, this is virtually risk free if the government is considered to be trustworthy. Some feed-in schemes only cover energy that is delivered to the grid, whereas other schemes also cover auto-production (using generated energy for own purposes).

'Net-metering' electricity producers that use (part of) the production for own consumption can use so-called 'smart meters'. A smart meter keeps track of the electricity supplied to the grid and the electricity taken from the grid. The owner of the smart meter will need to settle the net demand and supply from and to the grid (consumption and production may not coincide in time) and thereby level out the energy bill. This has an advantage, since the buying price of energy (which may include taxes) is usually higher than the selling price. For example, an individual who has a solar PV installation will weigh the costs of his installation against the kWh he no longer needs to buy (e.g. 20 ct/kWh), while a grid supplier needs to calculate with the selling price (e.g. 5 ct/kWh). Thus, for individuals who produce their own electricity, the financial 'break even' point is closer than for grid suppliers. Note that foregone taxes are effectively an additional subsidy for auto-producers. Moreover, net-metering may require an adjustment of legislation. Smart meters may or may not be used in conjunction with a feed-in scheme.

To avoid operational overhead for the government, feed-in schemes do not look at specific projects and real costs, but instead use cost estimates per category. As a result, within a category some initiatives may be economically viable whereas others are not. To build a viable business model based on a feed-in scheme, the investor thus has to undertake a careful assessment of the project economics taking into consideration e.g. climate conditions for solar PV or heat pumps, technology costs, and fuel prices, e.g. prices of biomass for a biomass boiler.

The main advantage of a feed-in based business model is that it has a predictable and stable long-term cash flow from a credit-worthy counterpart (Gifford et al., 2011). Investors may combine the use of a feed-in scheme with other available support mechanisms such as soft loans or fiscal incentives to improve the financing conditions.

Many RE technologies require high up-front investments and when this poses a barrier for investors, a part of the feed-in tariff may be made available as investment subsidy.

The financial structure of a feed-in based business model differs between initiatives that 1) produce more energy than needed for own consumption, and initiatives that will 2) need to buy additional energy from the market. The two situations below are simplified examples to indicate where the energy and monetary streams flow under a tariff and a premium situation, and under production below and above own use.

Figure 4.14 shows how a building owner generates renewable energy, first for own use and the excess is supplied to the grid. In a feed-in tariff situation, the building owner gets a fixed tariff for the electricity injected to the grid, i.e. supplied to the network operator (i.e. energy does not enter the market). Note that the use of certificates and the use of a smart meter are optional.

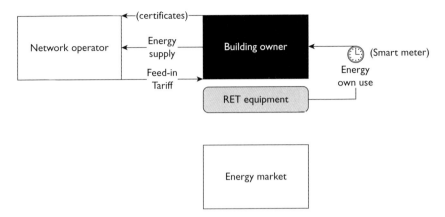

Figure 4.14 Schematic depiction of a business model based on a feed-in tariff and production exceeding own use, smart meter optional

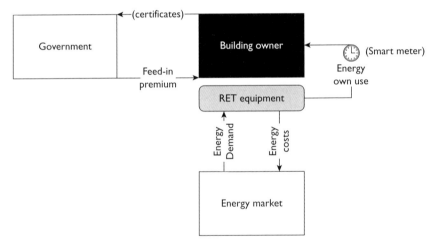

Figure 4.15 Schematic depiction of a business model based on a feed-in premium and production less than own use, smart meter optional

Figure 4.15 shows how a building owner generates renewable energy, but needs to purchase additional energy from the market to meet total demand. Note that the premium is based on certificates from the production side, and the use of a smart meter, which is still possible.

The basic concept of a feed-in scheme is standard, but the implementation can vary on, inter alia, the following parameters:

• Choice of categories by technology, size and use: how narrowly (or broadly) defined are the eligible technologies? Do categories differentiate by size, and take economies of scale into account?

- Volume and price limitations: is there a maximum (annual) budget beyond which feed-in contracts are no longer issued, and in case of a variable premium (such as in the Dutch SDE scheme) are there limits to the remuneration that the producer can claim?
- Duration: what is the duration of the guaranteed feed-in period, and how does it relate to the economic and technical lifetime of the installation?
- Legal issues and requirements: can a feed-in contract be passed on to a new owner of the installation, and under which conditions?

Existing markets and policy context

Feed-in schemes are currently the most commonly used incentive tools for renewable energy production worldwide. As of 2011, almost 50 countries had implemented a feed-in remuneration scheme (see Table 4.1) (REN21, 2011). The German EEG has been among the most well-known and effective feed-in schemes and this success can be largely attributed to its policy stability (Lensink et al., 2007). Other notable feed-in schemes that are relevant (but not necessarily limited) to the built environment include the Dutch MEP/ SDE (AgentschapNL, 2011), the German MAP programme (BMU, 2011) and (part of) the UK Low carbon buildings programme. There is a lot of experience with feed-in schemes for electricity, but little for heat. The UK Renewable Heat Incentive (RHI) is the first scheme for renewable heat (DECC, 2011b).

Feed-in tariffs have been the driving force behind the uptake of solar PV in the buildings sector in recent years.

Table 4.1 Countries with a feed-in remuneration scheme (Note, that not all of the feed-in remuneration schemes in the table extend to RET in the built environment)

	Countries with a feed-in remuneration scheme
High Income Countries	Austria, Canada, Croatia, Cyprus, Czech Republic, Denmark, Estonia, France, Germany, Greece, Hungary, Ireland, Israel, Italy, Japan, Latvia, Luxembourg, Netherlands, Portugal, Slovakia, Slovenia, Korea Republic, Spain, Switzerland, UK
Upper Middle Income Countries	Albania, Algeria, Argentina, Bulgaria, Dominican Republic, Lithuania, Serbia, South Africa, Turkey, Uruguay
Lower Middle Income Countries	Armenia, China, Ecuador, India, Mongolia, Nicaragua, Philippines, Syria, Thailand, Ukraine
Low Income Countries	Kenya, Tanzania, Uganda

(Source: REN21, 2011)

SWOT analysis

STRENGTHS

For the building owner/investor:

- A feed-in tariff can cover the financial gap between the generation costs of using renewable energy versus using conventional (fossil) energy.
- A feed-in agreement assures the producer of a predictable long-term source of income from a usually reliable counterpart (government or network operator), for the duration of typically 8–20 years. This can significantly reduce the financial risk of the project.
- Feed-in schemes are generally transparent and relatively simple, although the number of categories and eligibility criteria may grow over time.
- There is an incentive for entrepreneurs to find low cost projects, that 'outperform' the benchmark on which the category is based, e.g. at locations with very favourable climatic conditions or if the entrepreneur has access to comparably cheap supply of biomass. The scheme thus leaves room for profit by 'smart' entrepreneurs with an above average business case.

For government:

- A feed-in tariff can differentiate to meet the (cost recovery) needs of specific categories and actors.

WEAKNESSES

- Tariffs are based on cost estimates at a certain moment before payment. The actual implementation of an initiative that applies for the tariff can be up to years later. If in that time the real costs have increased, the rates are too low to make a solid business case. This is a risk for the producer in the project development stage. If real costs have decreased, the feed-in scheme allows for windfall profit for investors, decreasing the scheme's cost-effectiveness and potentially leading to political opposition.
- The administrative costs of metering may be significant compared to the total income from the feed-in scheme. This is especially the case for metering renewable heat production, because additional heat metering is needed that is more expensive then electricity metering. Solutions to this include using proxies and estimates, and using smart meters.

OPPORTUNITIES

- Because of the ability to differentiate, government can use feed-in tariffs to support specific technologies or markets with the aim of technology learning or market creation. Practical examples of a boost in market development include the solar industry, which developed strongly in recent years backed by consistent (feed-in) support from countries such as Germany and Spain.

- Feed-in schemes can provide a basis for developing an energy service company (ESCO), where the owner of a property outsources installation and operation to the ESCO, as well as the claim to the feed-in support. Thus feed-in schemes can potentially drive the deployment of RET in ESCO models.

THREATS

- The uptake of available feed-in schemes may be hindered by the fact that individual property owners may favour shorter payback times than feed-

Strengths
- Cover additional costs of RE
- Provides predictable long-term income for (existing) investors
- Transparent and simple
- Differentiates between
- technologies and actors
- Provides incentive for 'smart' entrepreneurs

Weaknesses
- Tariffs may deviate from actual costs over time
- Administrative costs for small producers may be high
- Feed-in schemes are based on long payback times

Opportunities
- Can help boost development of specific markets and technologies
- Provides favourable conditions fordeployment of RET in ESCOs
- Can be combined with other policy support

Threats
- Tariff setting requires insight in generation costs
- Exposure to policy decision making introduces uncertainty for supppliers and installers
- Decreasing public support in cases where feed-in tariffs are causing increases in electricity prices

Figure 4.16 Business model based on feed-in remuneration – summary of the SWOT analysis

in schemes provide (typically 8–20 years). The property owners may consequently favour direct investment subsidies over longer term compensation and may not make use of an available feed-in scheme.

- Budgets and tariffs per category are subject to policy (and political) decision making, which may lead to a certain level of unpredictability for suppliers and the installation sector. In years with low or even insufficient tariffs or budgets, the market may experience a sharp decline in demand.
- A successful feed-in tariff scheme may come at a price. Depending on the way it is financed, it either puts an increased burden on the energy consumers or, in the case of government funding, it may require significant use of public funds.

Discussion and conclusions

A feed-in scheme can be a solid basis for a business model. It provides a predictable, long-term income for investors, either in the form of a *tariff* (i.e. preferential price) or in the form of a *premium* (i.e. bonus). It can be used for every technology and market segment, but because the generation costs form the basis of the tariff levels, it works best when eligible installations do not vary widely in costs and a representative (fictional) reference cost calculation can be made. This makes it less useful for innovative and/or unique initiatives. Experience across Europe has shown that feed-in schemes are effective in developing markets, and boosting the use of RE both in and outside the built environment (REN21, 2011).

Considerations for building owners/investors

The strength of feed-in based business models is the long-term predictability of the income provided by it once a project has started. The weakness of a feed-in based system is that tariffs set may deviate (over time) from the real costs. Investors that plan multiple investments over time do require trust in government to keep the stability of the feed-in system, which may be difficult if the government allows itself too much flexibility and changes from year to year. Frequent changes and (unintentionally) low tariffs can slow down or even threaten market development.

Consideration for governments

Because of the flexibility in choosing categories and tariffs, government can use a feed-in scheme to stimulate private sector investments into specific technologies or niche markets. The flexibility for policy makers also presents some challenges. Unpredictable changes may have severe effects on the market in which suppliers and installers operate. In years where tariffs or budgets turn out to be too low, demand for new installations may plummet.

On the other hand, when tariffs are too high, deployment levels and the required support budget can become very high. Policy makers are commended not to use the flexibility of the feed-in scheme too much (beyond adjusting tariffs downwards as technology costs decrease), as the main success factor of existing schemes is their stability.

A feed-in tariff is a policy instrument that targets a specific financial barrier: that the generation costs of renewable energy are higher than of conventional, fossil fuel based energy. In reality, there may be more barriers that prohibit implementation, such as high up-front capital costs. Policy makers should therefore consider supporting RE in the built environment with a mix of different instruments, including soft loans, fiscal arrangements, and investment subsidies or grants. Feed-in schemes can work effectively alongside other policies such as fiscal benefits and investment subsidies.

Developing properties certified with a green building label

Description

'Green' building certification systems assess a building's performance according to environmental and wider sustainability criteria and provide proof that the building confirms to a certain sustainability standard.[8]

In this business model a property developer or architect designs and builds buildings certified according to a voluntary 'green'[9] certification scheme, expecting to realise a sales price premium compared to conventional buildings. This premium should compensate for the additional costs related to the 'green' features of the building, and for the costs of the certification. Drivers for an increasing demand for certified buildings include:

- Corporate social responsibility (CSR) considerations of corporations, for whom 'green' buildings are part of their 'green' image
- Reduced operating costs of 'green' buildings
- Enhanced levels of comfort for building users, which in commercial buildings may lead to higher productivity and less sick leave
- Regulation which mandates 'green' certification, for example for public buildings, and turns voluntary schemes into mandatory ones.

Most 'green' building certification systems cover a range of environmental and broader sustainability criteria related to energy and water use, indoor environment, and materials used, some systems also include criteria on functionality and comfort, economic questions and innovation (Nelson et al., 2010). Normally, a building must fulfil most of the criteria set by the certification systems. Most programmes include different levels of certification, for example Certified, Silver, Gold and Platinum for the US 'Leadership in Energy and Environmental Design' (LEED) standard.

There are a variety of voluntary certification systems globally. The most widely used are the US Green Building Council's LEED standards and the UK based 'Building Research Establishment Environmental Assessment' (BREEAM). There are also schemes which focus exclusively on energy related criteria, e.g. the US and Canadian Energy Star label for buildings, the German 'Passive house' standard and the Swiss 'Minergie' standard. In addition to certification systems there are also building rating systems, which do not issue a formal certificate. These rating systems support project developers by setting clear standards on what constitutes a green building. As rating a building is cheaper than undergoing a formal certification process, rating systems are frequently used for residential buildings (Nelson et al., 2010).

Market segments

Certification can be done in all market segments, i.e. for new and for existing buildings, for commercial, residential and public properties, and for rented and owner-occupied properties. This business model focuses on the sale of certified new developments (or on certification and sale after renovation). Appendix A.11 describes three examples of new LEED certified buildings.

'Green' building certification often targets the top-end of the real estate market. High-end office properties located in central business districts of large cities, corporate headquarters of multi-national companies but also newly built public buildings are typical examples of buildings designed and built according to the highest level of certification. A reason for the focus on the top-end of the market may be the fact that demand for certified buildings is often driven by non-financial reasons such as corporate reputation.

Originally, most certification systems have focused on certification of new properties. Today, most certified 'green' buildings space exists in commercial buildings, and a significant proportion of the certification is in existing buildings. For residential buildings, the focus continues to be on new properties (Pike research, 2010).

Applicable technologies

Which technologies can be included in business models based on 'green' building labels depends on the certification scheme: all 'green' certification schemes include energy efficiency measures. Under both the LEED and BREEAM certification systems, the installation of on-site renewable energy technologies is evaluated positively and contributes to the performance rating. (For examples of RET and EE measures applied under LEED, see also the case studies in Appendix A.11.) The 2008 BREEAM standard for new office building requires for example at a minimum a feasibility study for the integration of low or zero carbon technologies into a new office building.[10] However, it does not require the actual implementation of such technologies.

Actors

Directly involved actors are the property developer who designs and builds buildings certified according to the 'green' label and his clients who are willing to pay a premium for buying or renting certified property. Note that 'green' certification is also possible for existing buildings, but this business model analysis focuses on new builds. Other involved actors are the institution which develops and manages the certification system. Mostly these institutions are non-governmental: LEED is administered by the US Green Building Council (USGBC), BREAM is operated by BRE (Building Research Establishment), the Canadian R-2000 standard is administered by Natural Resources Canada (NRCan) and the SBTool was developed at the International Initiative for a Sustainable Built Environment (iiSBE). There may also be involvement by government institutions in the development and management of certification systems: BREAM was initially developed and marketed by the British government before the scheme was privatised (Nelson et al., 2010). The German Sustainable Building Certificate is run by the German Ministry of Housing.

Organisational and financial structure

This business model requires demand for buildings built according to above-average environmental ('green') standards. Certification is used to underpin and prove the building's environmental qualities. The design and construction of certified 'green' buildings is generally more expensive than of similar developments without improved environmental properties. In addition, there are costs related to the certification itself. The cost for a combined design and construction review of a new building of 4.650 m^2 or less applying for LEED certification is for example around USD 2,500.[11] A property developer must thus be able to cover the additional costs via a premium to the sales price of the property (see Figure 4.17).

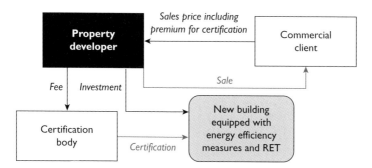

Figure 4.17 Schematic depiction of the business case of developing a property certified according to a green label

A number of recent studies show that certified 'green' buildings indeed have higher property values. A US study finds 13.5 per cent higher market values per square foot for Energy Star certified buildings than for non-certified ones (Pivo & Fisher, 2009). Other US studies find sales price premiums of between 5 per cent and 64 per cent for certified 'green' buildings compared to the non-certified samples (Fuerst & McAllister, 2008, 2011; Miller et al., 2008; Eichholtz et al., 2008). However, the large variation in this data also shows that there is still significant uncertainty regarding the real price premium of certified buildings, which may also depend on the level of certification achieved. Another open question is if sales price premiums are sustainable across property market cycles, i.e. if premiums persist during market slumps.

Similarly, there is a large variability regarding the estimates of additional costs to design and build a 'green' building. Some studies indicate that green buildings are not necessarily much more expensive than less efficient buildings. Kats et al. (2008) examined 146 green buildings in the US and found that the median of additional costs for the green aspects of the buildings was just 2 per cent. The study also found that energy savings alone make green buildings cost effective. These savings outweigh the initial cost premium in most green buildings. The net present value of 20 years of energy savings in a typical 'green' office ranges from USD 75/m^2 to USD 140/m^2, more than the average additional cost of USD 32/m^2 to USD 86/m^2 for building 'green'. In roughly 50 per cent of 'green' buildings in the study's data set the initial 'green premium' is paid back by energy and water savings in five years or less (see also Mathiessen & Morris, 2004).

Existing markets and policy context

So far, voluntary 'green' building certification systems have gained most traction in the US and the UK, which have the largest number of certified buildings and are home to the two largest voluntary certification schemes, LEED and BREEAMS. In 2010, the US Green Building Council announced that it had achieved 1 billion square feet (about 93 million m^2) of LEED certified commercial building space.[12] BREEAM has achieved 200,000 certified buildings and over a million registered for assessment since it was first launched in 1990.[13] In Canada, the voluntary standard R-2000,[14] which was developed in partnership with Canada's residential construction industry, and the ENERGY STAR for New Homes programme[15] have contributed to increasing awareness of energy issues in the residential sector and have also influenced the uptake of energy provisions in building codes.

Voluntary building certification is less widespread in continental Europe. Italy and Spain, for example, use the certification tools Protocollo ITACA and VERDE respectively which are based on the Sustainable Building Tool (SBTool) developed in Canada in the 1990s. Germany has introduced a building certification system only in 2009, the German Sustainable Building certificate, which is one of the most comprehensive certification systems worldwide, as it includes a wide range of criteria including economic and social ones (Nelson et al., 2010). Figure 4.18 shows which countries have voluntary 'green' building certification programmes.

Generally it seems that when comparing different OECD countries, countries with strict building standards and strong general environmental regulation have shorter histories of 'green' building certification systems than countries with less strict standards and that certification systems are less widespread in the countries with strong general environmental regulation (Nelson, 2008). An example for this hypothesis would be Germany, which has relatively strict building regulations, but which only recently introduced a certification system. Also, Scandinavian countries, which have a long history of environmental regulation, have so far focused mostly on 'green' rating systems. In the US, where environmental building standards tend to be lower, certification is much more common (Nelson, 2008). One reason for this may be that in countries with stricter existing regulation, there is little benefit in awarding certification to new buildings which may not be significantly 'greener' than the average new building. In addition, in countries with less strict existing building regulations, a certification system is likely to be less demanding than in countries with stricter regulation (Nelson, 2008).

There is a certain overlap between the voluntary 'green' certification programmes in Europe and energy performance labelling as mandated by the EU Directive on Energy Performance of Buildings (EPBD). In the EU, it is expected that the EPBD will be the main driver for moving the market as a whole, especially for new buildings. In May 2010, the European Parliament and Commission adopted the recast of the directive, which was to be transposed into national law by July 2012. The recast of the directive mandates member states to ensure that by the end of 2018 all new buildings occupied and owned by public authorities are nearly zero-energy buildings and that by the end of 2020 all new buildings achieve nearly zero-energy status. The directive further specifies that in these nearly zero-energy buildings, a very significant amount of the energy still required should be produced from renewable sources, including energy from renewable sources produced on-site or nearby (EC, 2010). Outside of Europe, similar developments are ongoing in other OECD countries. The government of Japan is for example considering to adopt a target for zero-energy buildings by 2030 (EC, 2009b).

Already under the 2002 version of the directive, member states are required to ensure that buildings have an energy performance certificate which is to be made available to the new owner or tenant when buildings are constructed, sold or rented out (EC, 2002). Since January 2006, certification is gradually being introduced in the EU member states for different types of

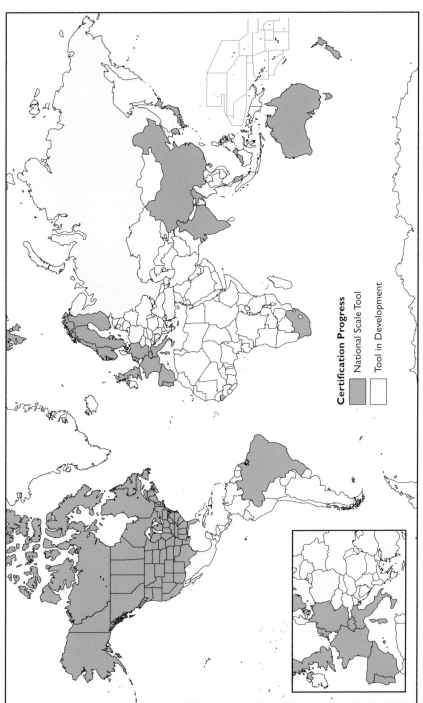

Figure 4.18 Countries with 'green' building certification programmes (source: Pike research, 2010)

Certification Progress

National Scale Tool

Tool in Development

buildings. Countries were required to implement mandatory energy performance certification of new and existing buildings, along with periodic certification of public buildings by 1 January 2009 at the latest. For public buildings, the energy performance certificate needs to be publicly displayed.

As the EPBD is a framework directive, there is considerable room for country specific implementation. Thus the certification schemes vary across EU countries, for example with regard to the moment when the prospective buyer/tenant is given the Energy Performance certificate. Currently, many member states still allow that the prospective buyer/tenant receives the energy performance certificate only at the end of the transaction, e.g. at the notary, but not during the period when the decision making process is still ongoing. Only with the recast of the EPBD will member states be required to ensure that key energy performance characteristics are already communicated while advertising the building for rent or sale.

Currently, little is known on the real impact of energy performance certificates on owners' decision making for buying, renting or renovating a building. Even a link between the energy efficiency of a building as stated in the EPC and its price on the market has not been proved yet.

The energy performance certificates in the EU exist in parallel to the voluntary certification schemes which mostly cover a broad range of environmental or wider sustainability criteria.

SWOT analysis

STRENGTHS

For the architect and property developer:

- Ability to get a premium price for property certified according to a 'green' standard.
- Designing and building certified 'green' buildings may strengthen an architect's/property developer's competitive position in an environment of increasing demand for 'green' buildings.
- Building certification can be used for marketing/image building.

For the buyer:

- Certification gives the buyer certainty of environmental features of the building. The owner can expect lower operating costs and additional benefits such as increased comfort for building users.
- Building certification can be used for marketing/image building.

WEAKNESSES

For the property developer and buyer:

- Undergoing the process of certifying a building carries relatively high transaction costs.

General:

- Certification schemes generally focus on the high-end of the real estate market, and are unlikely to drive a transformation of the broader market. In the EU, it is, for example, expected that the most important driver for a transformation of the market will be stricter government regulation. However, in some other markets which do not yet have strict building standards, building certification may be the first step towards regulation (see opportunities).
- Green certification schemes do not explicitly promote the use of RET in buildings. Higher levels of certification generally have higher requirements in terms of minimum standards that a building must meet, including stronger incentives for the inclusion of RET. However, in most schemes it's possible to achieve a high level of certification without use of RET.[16]

OPPORTUNITIES

- Increased awareness of the benefits of 'green' buildings may increase demand for building certification and for 'green' buildings in general.
- Increased knowledge and harmonisation of certification system is also expected to increase demand. In many European countries, building certification systems have only recently been introduced and are growing from a low basis. In countries where building certification has already become more widespread, rapid growth is taking place.
- In places where no strict building standards exist yet, certification may be the first steps in moving towards regulation at least for certain segments of the market. Singapore, for example, introduced a voluntary 'green' building certification scheme in 2005, the BCA Green Mark standard. Three years later, the basic level of the scheme became the mandatory standard for new buildings and retrofits with floor area exceeding 2,000 square meters.[17] In the UK, since 2008 any public healthcare buildings have been required to meet at least BREEAM level 'Excellent', all major new school buildings and refurbishments need to reach BREEAM level 'Very Good'.[18]

THREATS

- Where general building standards are high, it is expected to be more difficult to convince potential buyers of the added value of 'green' building certification.
- There are lingering misconceptions on the additional price of 'green' buildings among players in the property development value chain which hinder the increased uptake of certified 'green' buildings. A survey of building professionals showed that industry actors on average estimated the cost premium for 'green' buildings to be around 17 per cent whilst it is less than 10 per cent in most countries (WBCSD, 2007).
- There is still a lack of comprehensive and transparent data demonstrating that certified 'green' buildings command higher sales prices, and a lack of studies on the overall benefits of 'green' buildings. Studies such as the ones quoted above focus mostly on the situation in the US; for Europe no comprehensive data is available, yet.

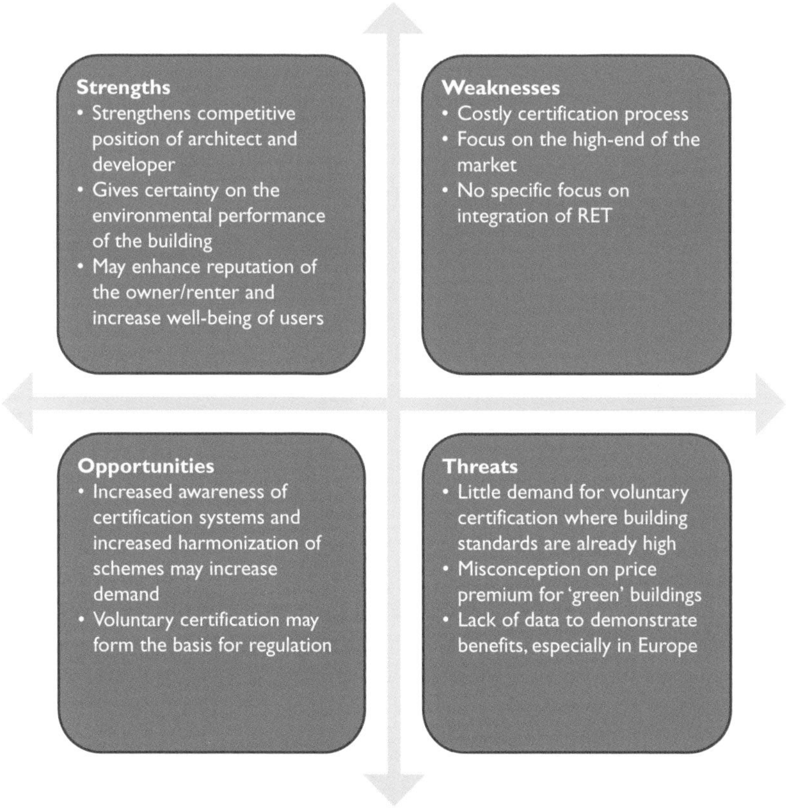

Strengths
- Strengthens competitive position of architect and developer
- Gives certainty on the environmental performance of the building
- May enhance reputation of the owner/renter and increase well-being of users

Weaknesses
- Costly certification process
- Focus on the high-end of the market
- No specific focus on integration of RET

Opportunities
- Increased awareness of certification systems and increased harmonization of schemes may increase demand
- Voluntary certification may form the basis for regulation

Threats
- Little demand for voluntary certification where building standards are already high
- Misconception on price premium for 'green' buildings
- Lack of data to demonstrate benefits, especially in Europe

Figure 4.19 Voluntary 'green' building certification – summary of the SWOT analysis

Discussion and conclusions

Voluntary 'green' building certification can play a role in driving the transformation of the property market towards becoming more environmentally friendly and energy efficient, especially in countries which do not (yet) have strict mandatory building codes. In Canada it is, for example, assumed that existing voluntary 'green' building certification programmes have influenced the uptake of energy provisions in building codes.

Certification enhances transparency on environmental characteristics of buildings and provides the owners and users of the building with some certainty on environmental performance, including an indication of operating costs for energy and water use. There are additional benefits to the users of green buildings such as enhancing corporate reputation and increasing well-being and productivity of building users. With increased awareness of the feasibility and benefits of voluntary certification, demand for certified buildings is expected to grow. The US and the UK in particular have seen significant growth in certified building space in the past years. Growing demand for certified buildings provides opportunities for architects and property developers. Recent studies in the US indicate that property values of certified buildings are indeed higher than of comparable non-certified ones. Other studies indicate that certified buildings may not be much more expensive to design and construct.

Regarding energy related aspects, the primary focus of most certification schemes is on the integration of energy efficiency measures into buildings. However, on-site production of renewable energy is also part of the most common schemes, especially for higher levels of certification.

Government can encourage certification schemes by applying them to public buildings or even making them mandatory for certain types of (public) buildings as is happening currently in the UK.[19] Moreover, government use of certification schemes raises awareness of the scheme. However, voluntary certification generally tends to focus on the high-end of the real estate market, and by itself is unlikely to drive a transformation of the whole market towards more environmentally friendly and energy efficient practices. In the EU, it is rather expected that the recast of the EU Directive on Energy Performance of Buildings (EPBD) will be the main driver for moving the market as a whole, especially for new buildings.

For owners and users of certified buildings, existing studies, mostly based on data from the US market, show operating cost savings and additional benefits such as increasing well-being and productivity of building users (see above). However, especially for the European context there are no major studies yet, demonstrating the enhanced financial performance of green buildings over conventional ones (Nelson et al., 2010). Certification bodies and property developers and owners with a large portfolio of certified green buildings could take a proactive role in collecting data and initiating such studies in order to increase confidence of investors into the concept. Furthermore, governments and certification bodies could work together to ensure that the requirements for voluntary and mandatory systems are harmonised. The latter is especially important in markets where voluntary certification has already reached a high market share in order to ensure that competing and incompatible systems do not cause confusion with investors and other market actors.

Building owner profiting from rent increases after the implementation of energy efficiency measures

Description

In this business model, building owners (who do not occupy a building themselves) or housing corporations benefit from additional revenue opportunities after having undertaken investments in RET and EE: they may be allowed to charge a higher rent from their tenants after the renovation because these benefit from lower energy costs. This helps overcome the barrier of split incentives, i.e. the lack of incentives to realise building improvements when owner and occupant are different parties.

This business model is based on regulation that allows such rent increases and is being introduced in a number of countries (see below). Such regulation is possible in situations where regulation on maximum rents and/or maximum allowable rent increases exist. This is usually the case in the social housing sector, but such regulation may also exist in the wider residential rental sector where buildings are owned by private persons or property companies.

Market segments

This is applicable for renter-occupied residential buildings in jurisdictions where the rental sector is regulated through determined maximum levels of rent or maximum allowable rent increases. It is mostly specifically applied in the social housing sector which is usually protected by tenants' law and where there is a system for determining the level of rent based on a set of criteria.

The absolute volume of buildings in the social housing sector in the EU is significant. The relative share of social housing in the total building stock in EU countries is estimated to be on average 13 per cent. In most EU countries the social housing sector is regulated (OTB, 2010).

The rental sector is also significant in size outside of Europe, for example in the US and Japan, where the majority of apartments are rented (WBCSD, 2007).

Applicable technologies

Theoretically both EE measures and installation of RET could be undertaken under this scheme. Practically, the regulation is expected to be mostly used for energy efficiency measures because they are usually more cost effective.

Actors

Directly involved actors are property owners (housing corporations, individuals, corporate or institutional investors) and tenants. The business model also involves governments which set the rental regulations and other actors involved in the building sector such as installers of energy efficiency measures and energy auditors.

Organisational and financial structure

A building owner in a regulatory environment that allows higher rents for buildings with higher energy performance decides to undertake improvements to the energy performance of his property. To compensate for his investment, he increases the rent of his tenants who profit from lower energy costs. In doing so, the building owner aims at recovering his investment through the higher rents over a reasonable period of time (see Figure 4.20).

Regulation in many countries determines the maximum level of rent that can be charged or maximum allowable rent increases. To enable the building owner to pass through (part of) his costs, a change in regulation is required. Especially in the social housing sector, government generally aims to ensure that living expenses for tenants decrease or at least do not increase. Thus the increase in rent should be lower than the energy cost savings, which limits the choice of available energy efficiency measures that an owner can cost-effectively undertake under the scheme.

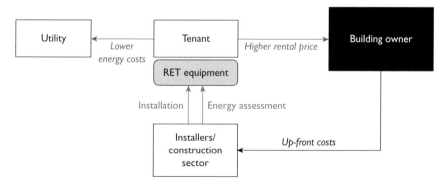

Figure 4.20 Schematic depiction of the business model

For the proposed scheme in the Netherlands, the example in Appendix A.6 illustrates cost implications for renters and landlords of the approach for determining rents and describes the scheme in detail. Note that in the Dutch example the new regulation should guarantee that total costs for the tenants do not increase.

Existing markets and policy context

The regulation of the rental market differs widely across Europe. The same applies specifically for social housing and its regulations, although most social housing is managed by social housing corporations (Fresh project, 2011).

Regulations that help overcome the split incentive barrier in the rented sector are not common in Europe, although awareness of the need for such schemes is growing. This was demonstrated by a survey among real estate professionals and property owners' associations in many EU member states (UIPI & CEPI, 2010). Few countries have actually adapted their rent regulations to allow for increased rents after renovation (UIPI, 2010; IEE workshop, 2011). European countries with existing policy to address the split incentive issue in the rental market are the Netherlands, France, Germany, the UK, Italy and Sweden (IEE workshop, 2011; UIPI & CEPI, 2010). In the Netherlands, the tenants' law, more specifically the rental price evaluation system for social housing, is expected to be adapted in the course of 2011 to allow rental price increases by housing corporations for energy improvements made (see Appendix A.6) (Aedes, 2011).

In France, the tenants' law was adapted in 2009 enabling landlords that realise energy improvements to share the energy saving benefits with their tenants. A specific feature of the regulation, similar to the one in the Netherlands, is that a tenant has to give consent to the landlord to undertake the renovation. Furthermore, benefits to the landlord cannot exceed half of the energy cost savings (UIPI & CEPI, 2010).

In Germany, there is a green rent index to reward energy improvements. This index is not widely used though, as the opportunities for landlords to increase rental prices for investments into energy improvement are limited. Moreover, RET are not eligible for the index yet (Nelson et al., 2010; UIPI & CEPI, 2010).

In the UK the national energy efficiency plan of the government ('Green Deal') aims to establish financing options ('Pay-as-you-save schemes') for tenants when landlords realise energy efficiency improvements to their house (UIPI & CEPI, 2010).

In Belgium, Austria and Bulgaria discussions are ongoing to address the split incentive problem (UIPI & CEPI, 2010; IEE workshop, 2011). And in the future, it is expected that more EU member states will introduce similar regulation driven by the proposed replacement of the EU Energy Service Directive by a new Directive on Energy Efficiency. The proposed new directive acknowledges split incentives as a barrier for energy efficiency. Article 15 of the proposed directive states that member states should take action to remove the split incentive between the owner and tenant, for example by means of providing incentives, establishing public funds for energy efficiency to which all qualified service providers should have access or adjusting legal and regulatory provisions (European Parliament, 2012). There is no indication that similar regulation exists outside of Europe.

In most cases, the business model requires supporting policies or services. For example the assessments of a building has to be done by an energy label or audit, which implies that the country already needs to have implemented the respective requirements set in the EU Energy Performance of Buildings Directive (EPBD), or have a similar system in place. In addition, the business model may be regulated by additional policy, such as specific rules protecting tenants. For the case of the Netherlands, the Dutch living expenses guarantee is described in Appendix A.6.

SWOT analysis

STRENGTHS

- Reduces the split incentive barrier in the rental sector because of the benefits for both tenants and landlords (Tigchelaar et al., 2011).
- Stimulates energy improvements for existing dwellings on a large scale, as large property owners, e.g. social housing corporations, frequently have sufficient access to capital, technical expertise and a long-term interest in maintaining their building stock. Housing corporations in the Netherlands for example have relatively good access to capital at attractive interest rates because of a guarantee fund by the government for social housing corporations, CFV[22] (ECN, 2011a).

WEAKNESSES

- The business model is only applicable for rented buildings in countries or regions where rents are regulated. This is usually the case in the social housing sector, but may also be the case in the private rental sector.
- The scope of the business model is limited, as in social housing the tenants are protected by tenants law which may need to be changed. Buildings are therefore only renovated when new tenants move in or existing tenants provide consent (as required in some cases). Moreover, the amount of energy savings that property owners are allowed to recover may be limited, as is, for example, the case in France (Fresh project, 2011). This reduces the incentive for renovations. In the Netherlands, it is estimated that the requirement for owner consent will limit the amount of renovations undertaken (see also Appendix A.6).

- The business model may primarily lead to energy efficiency improvements instead of renewable energy technology deployment, as the latter is frequently more complex and expensive. Existing policy schemes therefore mostly focus on energy efficiency and limit or leave out RET.
- Effective enforcement of the scheme requires an energy performance assessment (for example by labeling).[21]
- Some property owners and housing corporations may not have access to capital to invest in energy improvements of buildings.

OPPORTUNITIES

- It can provide a significant driver for energy improvements of the existing building stock, for which energy saving potential is the high.
- More EU countries are expected to introduce similar regulation as part of their efforts to implement the revised EU Energy Services Directive. Outside of Europe, little information is available on planned efforts.

THREATS

- The business model requires a change in regulation that covers the rental market. This may be a time consuming process, as there are potential conflicts of interests between renters' associations and property owners. In the Netherlands, the change of the rental price evaluation system was, for example, only realised after a political process that took three years.
- Rented buildings with better energy performance but higher rent may be perceived as less affordable by a tenant.

Discussion and conclusions

The business model is based on a change in legislation regulating the rental sector. Its attractiveness for the building owner directly depends on the details of the legislation, e.g. how much of the energy savings or of his up-front investment a building owner is allowed to recover. It is unlikely that being able to charge higher rents to tenants will be the sole driver for a property owner's decision to undertake renovation measures. However, the higher rents may still play a significant role in the decision. It is expected that in its current form the business model is mostly applied for the implementation of energy efficiency measures which are usually more cost-effective than RET. But theoretically the business model may also be applied for the implementation of RET, e.g. for the installation of a heat pump which reduces energy costs for the tenant.

There are only few new business models and innovative policy instruments which specifically address the barrier of split incentives. This implies that this business model, potentially supported by additional incentives, may play an

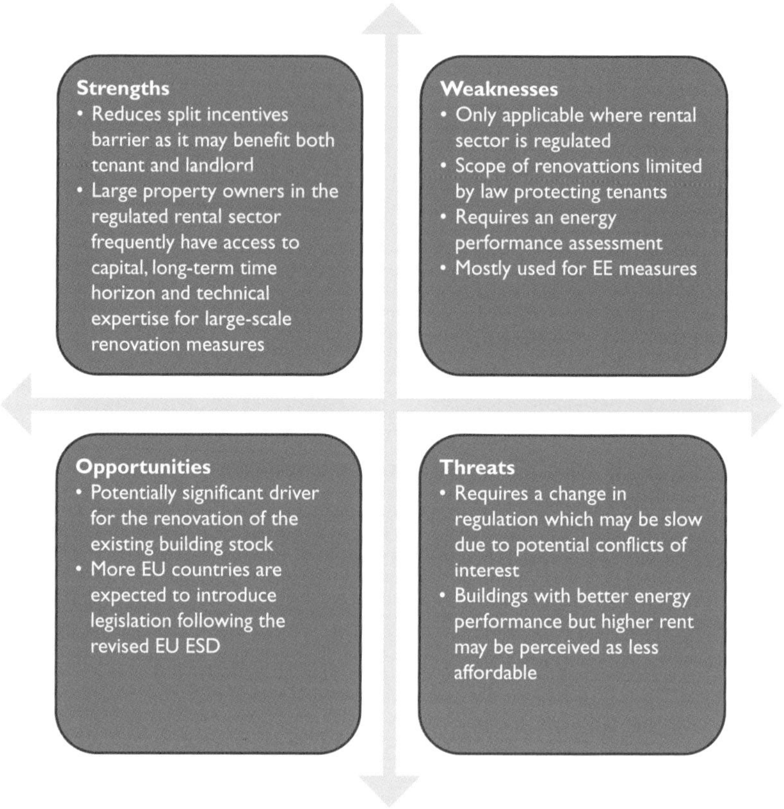

Strengths
- Reduces split incentives barrier as it may benefit both tenant and landlord
- Large property owners in the regulated rental sector frequently have access to capital, long-term time horizon and technical expertise for large-scale renovation measures

Weaknesses
- Only applicable where rental sector is regulated
- Scope of renovattions limited by law protecting tenants
- Requires an energy performance assessment
- Mostly used for EE measures

Opportunities
- Potentially significant driver for the renovation of the existing building stock
- More EU countries are expected to introduce legislation following the revised EU ESD

Threats
- Requires a change in regulation which may be slow due to potential conflicts of interest
- Buildings with better energy performance but higher rent may be perceived as less affordable

Figure 4.21 Business model based on higher rents after improving energy performance of a building – summary of the SWOT analysis

important role in catalysing energy improvements of the existing building stock in the large rental sector. The application of the business model is limited to countries or regions that have a regulated rental sector. However in the regulated rental sector mostly large property owners are active, such as social housing corporations which frequently have the long time horizon, access to capital and technical expertise required to plan and undertake renovation measures.

Business models based on new financing schemes

Property Assessed Clean Energy (PACE) financing

Description

Property Assessed Clean Energy (PACE) financing is a mechanism by which property owners finance RE and EE measures via an additional tax assessment[22] on their property. The property owners repay the 'assessment' over a period of 15 to 20 years through an increase[23] in their property tax bills (NREL, 2010). When a property changes ownership, the remaining debt is transferred with the property to the new owner. The concept is also referred to as 'energy financing districts' or 'tax-lien financing'.[24]

Market segments

The PACE business model can in principle be applied to all buildings for which the owner is eligible for property taxes. The model is relatively new, and current programmes in the US apply to owners of existing free-standing residential houses and commercial buildings.

Applicable technologies

Applicable technologies are RET, such as solar PV or solar thermal, efficient boilers, and energy efficiency improvements. Initial experiences with PACE financing suggest that due to the administrative costs associated with the mechanism, tax assessments (i.e. the total amount of the loan) should generally be at least USD 2,500 (US DoE, 2010a).

Actors

Directly involved actors are the building owner, who decides to install the RET/EE measures, and the city or local government who issue the loan and collect the repayments.

Organisational and financial structure

Figure 4.22 shows a typical organisational set-up of a PACE financing model, where the local government finances the programme, for example via issuing of municipal bonds. When a building owner decides to participate in the programme he needs to register and undergo certain eligibility checks, which are generally less thorough than when, for example, applying for a bank loan, because the debt will not stay with the owner but with the property. The

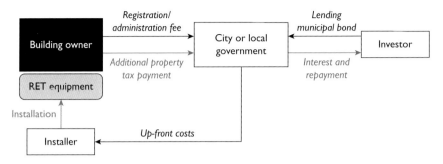

Figure 4.22 Schematic depiction of PACE financing for RET

check typically focuses on the land title and the property tax payment history (RAEL, 2009). The programme administration pays the installer directly, i.e. the up-front investment costs do not pass through the hands of the building owner. In return, an assessment is placed on the owner's property which is secured by a senior lien. Thus in the event of foreclosure (and forced sale), the local government will be paid back the PACE loan before any other claims against the property. If the owner sells the house before the end of the repayment period, the remaining debt, repayment obligation and the equipment are transferred with the property to the new owner.

The local tax agency acts as the collecting agent for the repayment. For a US$20,000 tax assessment at an interest rate of 6 per cent over 15 years, the annual repayment would for example amount to US$2,060. Generally, PACE financing instruments aim at structuring RE and EE measures in a way that the additional property tax payment is lower than the cost savings achieved, thus aiming at annual net cost savings for the building owner.

Assessing which measures are cost-effective requires at a minimum a rough energy audit of the property. Such services may be facilitated within the frame of the PACE financing programme, e.g. the administrative body can recommend service providers or may even offer energy audits themselves. These may be crucial for the programme to succeed, as building owners need to know in advance if the investment is cost-effective. The ownership of the RET system or EE technologies financed through a PACE financing programme lies with the property owner. Thus the property owner could legally be eligible for additional subsidies or incentives, e.g. a feed-in remuneration or tax benefits.

Whilst municipal bonds are the most typical way of financing a PACE programme, other options are possible, such as bank loans, general local government funds or existing revolving funds (see NREL, 2010 and RAEL, 2009). Depending on the type of financing used by local governments, interest rates for home owners may differ substantially. In the Berkeley First Program, for example, some applications to participate in the programme were withdrawn due to unattractive interest rates at 7.75 per cent (see case study in Appendix A.2). Two other early PACE financing programmes in Palm Desert, California and Boulder Country, Colorado offered interest rates of 6.68 per cent and 7 per cent respectively. On the other hand, a programme in Babylon, New York, using an existing municipal solid waste revolving fund, offered interest rates of only 3 per cent which was significantly below market rates (RAEL, 2009).

Existing markets and policy context

The PACE concept started in 2008 in California and is currently being implemented in various US counties and municipalities, and in a similar programme in Melbourne, Australia (see Appendix A.3 for a description of this programme in Melbourne). Under US law, as a prerequisite of a PACE programme, a state must establish legislation which enables local governments to create special assessment districts which consider RE and EE measures to be 'public goods'. In addition, the local government needs to establish regulation which create assessment zones and authorises the creation of liens on property. The local government also has to establish the funding process and organise the administration of the programme. In Australia, the introduction of the Environmental Upgrade Charge[25] for the city of Melbourne (see Appendix A.3), which is similar to the concept behind PACE financing, was also done on a state level.

At the end of 2010, 26 US states had passed legislation which allows local governments to create PACE programmes. There are active PACE programmes in California, Colorado, Wisconsin, New York and Maryland (Institute for Building Efficiency, 2010b).

The US Department of Energy offered support to PACE programmes in the form of technical assistance to local governments and through committing funds to communities which can be used to give low or no-interest loans for PACE programmes. However, as of early 2011 most PACE programmes are stalled because of a conflict about the senior liens placed on the property. Fannie Mae and Freddie Mac, which are the government entities guaranteeing more than half of US residential mortgages, rightfully fear that the senior liens on the property will impair their own claim in the case of default, as the property tax assessments typically need to be paid back first. In July 2010, the Federal Housing Finance Agency (FHFA) issued a statement instructing Fannie Mae and Freddie Mac to no longer underwrite mortgages for properties with a PACE assessment. As a consequence, a house owner now typically has to first pay back the property tax assessment before he can sell his home. This makes one of the major advantages of PACE financing obsolete and has stalled most programmes. Currently legal and legislative efforts are underway to restore PACE financing, but its future is still uncertain (Boland, 2010; PACENow, 2011; Woody, 2010). Commercial PACE programmes are not affected by the action by FHFA and continue to move ahead (US DoE, 2010b).

While the PACE concept has gained relatively widespread attention, its actual use is still limited. RAEL (2009) show that as of August 2009 in four of the early PACE programmes between 39 and 393 projects had been undertaken per programme with average project sizes between USD 7,100 and USD 36,000. Looking at purely commercial PACE programmes, as of early 2011 there were 4 programmes in operation and 9 in design. Within the 4 operational programmes, 71 projects of total financing of USD 9.69 million were approved (LBNL, 2011).

Outside of the US, the city of Melbourne in Australia introduced a so-called Environmental Upgrade Charge attached to the property. The programme aims to catalyse the retrofit of at least 1,200 non-residential buildings (LBNL, 2011). The programme is described in detail in Appendix A.3.

SWOT analysis

STRENGTHS

For the property owner:

- PACE financing overcomes the barrier of high up-front costs for the home owner.
- If the property tax assessment can be transferred together with the property to the new mortgage holder, the PACE model helps to overcome the hesitancy of home owners who may move house every 5 to 7 years to make long-term investments into RE and EE measures (NREL, 2010).
- If RE technologies are installed and contribute to a significant share of electricity demand, the property owner effectively fixes his energy costs for the next 15–20 years at the level of the additional property tax payment. This is attractive if electricity prices are expected to rise.
- PACE financing improves access to capital and allows for repayment terms of 15–20 years, much longer than typical home equity loans.
- PACE financing reduces transaction costs for the home owner as the programme is specifically set up to finance RET and EE measures (RAEL, 2009).
- Home owners may consider the local government to be a more trustworthy source of information on opportunities for RET and EE measures than for example industry organisations (Bailey & Broido Johnson, 2009). Similarly, investors may regard local government as a trustworthy partner.
- RET and EE measures undertaken with PACE financing are chosen based on having positive net annual cash flows. If the realised energy savings are indeed higher than the additional property tax payments, the

value of the property increases and the owner may realise a higher price when selling the house. If this is not the case, and realised energy savings are less than the additional property tax payments, there is a risk of a negative impact on the value of the property.

For the municipality:

- PACE financing generally allows a municipality to promote RET and energy efficiency with little direct costs to the government for financing such measures as the government does not need to use its own funds. The only actual costs for government are the administration costs of the programme (which may also be passed on to participating home owners).

WEAKNESSES

For the property owner:

- Even though other conditions may be more beneficial, interest rates for PACE financing are not necessarily lower than for alternative financing options.

For the municipality:

- Requires a change in legislation to establish the 'special assessment district'. In the US, the legislative change is the responsibility of states; in other countries responsibility may lie with a different level of government. For further information on current legal issues around PACE financing in the US, see also Nostrand (2011).
- Municipal governments may not have the right expertise to establish and run a PACE financing programme.
- Setting up and administering a PACE financing programme requires an administrative effort by the municipal government. For small municipalities this may be too costly, as they are not able to achieve economies of scale.

OPPORTUNITIES

- As municipalities gain more experience with the set-up of PACE financing programmes, lessons learned and best practices emerging from the first-mover programmes may be used to inform subsequent efforts and facilitate the set-up of new programmes. In the future, the PACE construction may pose opportunities to create regional programmes, thus lowering administrative costs for single municipalities (RAEL, 2009).
- PACE financing does not directly address the barrier of split incentives and does not specifically target rented buildings. However, in the US, tax

assessments qualify as a pass-through under so-called 'triple net lease' arrangements where the tenant agrees to pay all real estate taxes, building insurance, and maintenance (Institute for Building Efficiency, 2010a). Thus PACE financing can also be used to specifically target rented properties.

THREATS

- A major threat to residential PACE financing programmes in the US is the decision by the Federal Housing Finance Agency (FHFA) that instructs the mortgage reinsurers not to accept the PACE tax assessment when guaranteeing new mortgages. This makes it virtually impossible to sell houses with a PACE tax assessment, and has stalled residential PACE programmes. There is not solution for this issue, yet.
- Overall, the transferability of liabilities from PACE financing has not yet been proven on a large scale.
- PACE financing requires the approval of the mortgage holder and potentially other actors such as the mortgage re-insurers, which may be a threat to PACE programmes (see above). For commercial PACE financing programmes, several recent examples demonstrate that commercial building successfully obtained consent from mortgage holders, with a tax assessment-to-(building) value (lien-to-value or LTV) ratio of less than 1:10 (LBNL, 2011).
- With the recent economic recession, the financial situation of municipalities and local governments has been deteriorating in many countries. If municipalities are under strong budget pressure, they may not have the resources to set up and maintain a PACE financing programme. For many municipalities and local governments economic prospects are expected to continue to be difficult in the coming years due to decreasing local tax revenues and less support by national governments.

Discussion and conclusions

PACE financing is a relatively new concept: the first pilot programme was undertaken in 2008 in the city of Berkeley. It is attractive as it provides access to capital for property owners to invest in RET and EE measures, thus overcoming the barrier of high up-front costs. In addition, it incentivises long-term investments because the repayments of the special tax assessment may be done over 15 to 20 years and the lien on the property stays with the property when it is sold. PACE financing for residential properties suffered a serious draw-back when the US Federal Housing Finance Agency issued a statement that it would not accept mortgages with liens for PACE financing, implying that a property owner would first have to pay back the tax assessment before being able to sell his property. While this decision has put

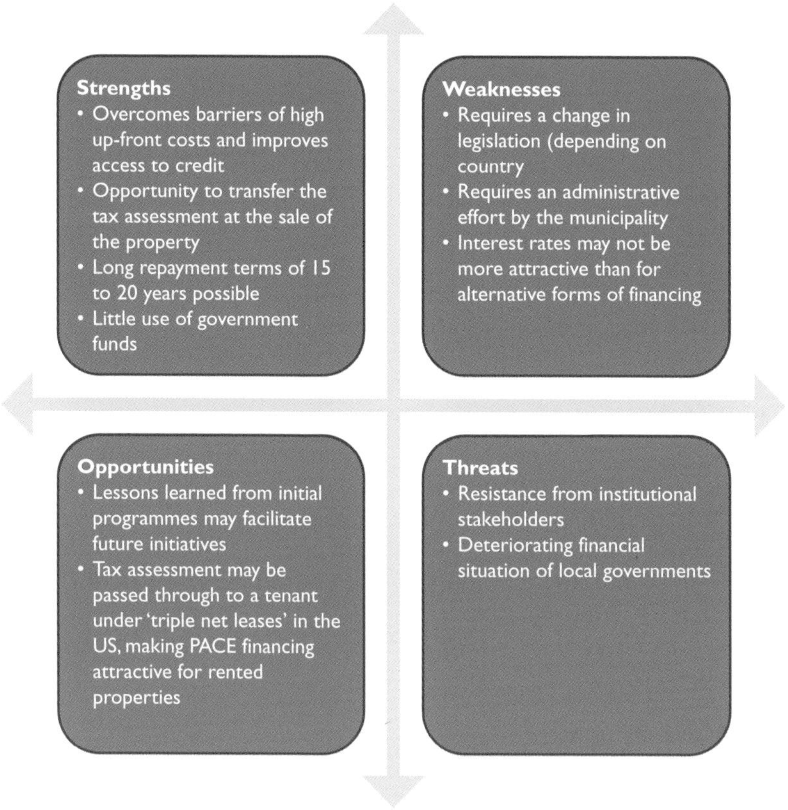

Strengths	Weaknesses
• Overcomes barriers of high up-front costs and improves access to credit • Opportunity to transfer the tax assessment at the sale of the property • Long repayment terms of 15 to 20 years possible • Little use of government funds	• Requires a change in legislation (depending on country • Requires an administrative effort by the municipality • Interest rates may not be more attractive than for alternative forms of financing
Opportunities	**Threats**
• Lessons learned from initial programmes may facilitate future initiatives • Tax assessment may be passed through to a tenant under 'triple net leases' in the US, making PACE financing attractive for rented properties	• Resistance from institutional stakeholders • Deteriorating financial situation of local governments

Figure 4.23 PACE financing – summary of the SWOT analysis

residential programmes to a halt as long as no solution for this issue is found, commercial programmes continue. For residential programmes legislative and legal efforts are underway to solve the outstanding issues. However, if no solution is found, it is unlikely that residential PACE programmes will continue.

The main prerequisite to enable PACE financing or a similar programme is a change in legislation that enables the creation of special tax assessment districts or similar arrangements, and the acceptance of all stakeholders (as indicated above). In the US and Australia, these legislative changes are undertaken on a state level. If other countries were interested in introducing such a concept, national or regional governments would have to explore if and how the concept could fit into the existing regulatory framework.

Considerations for local governments

Local governments which consider the introduction of a PACE financing programme are advised to first undertake an analysis of the potential market for PACE financing. Key points to consider include:

- The demographics of the area and insight in the split between owner-occupied and rented properties, the split between commercial and residential properties, income levels, and awareness and interest in RE and EE measures.
- The current state of the building stock, the climatic conditions, and current energy sources and prices for electricity and heat. These will determine which technologies would be appropriate for inclusion into a PACE financing programme.
- The current barriers to investing into RET and EE measures. PACE financing may overcome the barrier of high up-front costs and the hesitancy to make long-term investments, but it does not directly address the problem of split incentives, of low or negative returns on investments in RET and EE measures, or lack of awareness and information. If other barriers prevail, these need to be addressed first or in parallel with the PACE financing programme.
- The availability of existing financing options for property owners (RAEL, 2009).

In addition, the ability of the local government to secure funding for PACE financing (RAEL, 2009) is an important consideration when setting up a similar programme. Regarding the administrative overhead, the local government may consider if it wants to undertake the programme itself or rather partner with third parties (private or public) to lower the burden. Even though there is not much experience, existing PACE programmes may provide valuable input for the set-up of new initiatives.

Considerations for property owners

Property owners who have access to PACE financing should consider if financing conditions via PACE financing are attractive compared to other alternatives such as direct bank loans or own capital. It is also important to take into consideration that an investment which leads to annual cost savings, i.e. where the annual energy cost savings are higher than the additional tax payments, probably leads to an increase in the value of the property. On the other hand there is a risk that the property value decreases if the special tax assessment minus the energy cost savings create additional costs to a prospective buyer.

Box 4.1 Case in point: Berkeley FIRST – the first PACE financing programme

In 2008, the City of Berkeley launched the first ever PACE financing programme, called Berkeley Financing Initiative for Renewable and Solar Technology (FIRST). The small pilot programme focused exclusively on solar PV installations in order to keep the process simple. Funding was based on micro-bonds issued by the City of Berkeley. The programme provided financing to 13 projects for a total of USD 336,550 of financing. The Berkeley FIRST programme provided valuable insights into how PACE financing can be applied in practice (see Appendix A.2 for more details).

On-bill financing

Description

In this business model, utilities provide financing (i.e. a loan) for RE and EE measures. The building owner (or building user) repays this loan via a surcharge on its utility bill. Preferably the overall utility bill should still be lowered, though, because of the associated energy cost savings. It is possible to structure the programme in a way that the loan stays with the utility meter, and thus can be transferred to the new owner if the house is sold.

Market segments

On-bill financing targets mostly owners of free-standing residential homes and small commercial buildings who want to upgrade existing buildings. It is applicable for owner-occupied buildings, but can also work for renter-occupied buildings, as the concept may allow tenants to pay for (via the utility bill) and profit from energy efficiency improvements. In the US, the two utilities Electric Company (HECO) and Midwest Energy have pioneered on-bill financing programmes which specifically target renter-occupied buildings (see Johnson et al. (2011) for an evaluation of these programmes).

Applicable technologies

All RE and EE measures are theoretically possible. When applying on-bill financing, the aim is generally to generate annual cost savings for the building owner within the loan term which must be shorter than the useful life of the equipment. Thus, only cost-effective measures are financed. On-bill financing programmes are often combined with grants to enable a wider range of measures to be cost-effective. North American programmes usually finance EE measures via on-bill financing. Many programmes either offer a predetermined list of technology options or require an initial energy audit, after which generally also one or several options out of a predetermined list are suggested (Brown, 2009).

Actors

On-bill financing programmes are set-up by utilities. In addition, they involve building owners, who decide to initiate RE and EE measures through on-bill financing. Utilities are frequently able to finance the programmes themselves as they have sufficient equity capital and access to debt facilities. However, the utility may also rely on additional partners for financing, such as banks or government bodies, e.g. through revolving funds. Installers of RE equipment may be involved by partnering with the utility. Successful programmes are often characterised by strong partnerships between involved actors (Johnson et al., 2011; see also PROSOL case study in Appendix A.4 and the case study of Manitoba Hydro in Appendix A.10).

Organisational and financial structure

Figure 4.24 shows a simple model of how an on-bill financing programme may work. In this example a utility provides a loan to a building owner for the installation of RE technology. The building owner in turn repays the principal and interest via its utility bill.

In the example in Figure 4.24, the utility does not only administer the programme and collect payments via electricity bills, but it also finances the investments from its own capital. In the US, most on-bill financing programmes have been funded by a combination of sources such as the utility's owns funds, special charges levied from all clients with the aim to support RE and EE measures and government funds. Potentially, funding could also come from issuing bonds, or other public or private sources of capital (Fuller, 2008).

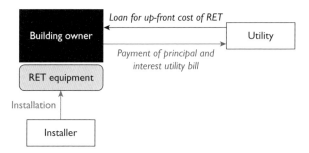

Figure 4.24 Schematic depiction of on-bill financing of RET

There are different types of on-bill financing programmes, which in the US are distinguished as *on-bill loan* and *on-bill tariff* programmes (Brown & Conover, 2009): With an *on-bill loan* programme, a personal loan is issued to the building owner, repaid as a line-item on the utility bill. However, it is legally not linked to the property or the utility meter.

In *on-bill tariff* programmes on the other hand, the building owner also repays the loan via the utility bill, but in this case it is considered an 'essential service' and part of tariff, that the utility charges its customers (Brown & Conover, 2009). The obligation for payments stays with the property and is transferred to the next owner in the case of sale of the property as with PACE financing. Brown and Conover (2009) provide examples for both types of programmes. They also provide an overview of interest rates used in their case study programmes in North America, where rates vary between 0 per cent and 8.5 per cent.

Generally, the target for the investments in RE and EE measures is to generate positive cash flows for the property owner. As a consequence, the repayment periods vary depending on the expected energy savings and the useful life of the installed measures (Brown & Conover, 2009). In the US and Canada, on-bill financing programmes for energy efficiency retrofits frequently have mid-length loan terms: for *on-bill loan* programmes, repayment periods are often set at around 5 years. For *on-bill tariffs*, where the loan is legally linked to the utility meter, the payback period of the loan is stretched to about 10 years (Brown & Conover, 2009).

Existing markets and policy context

On-bill financing programmes have been undertaken by various Canadian and US utilities, for example in New England and California. In North America, successful on-bill financing programmes date back over 10 years (Franklin Energy, 2011). The case study of the on-bill financing programme of the Canadian utility Manitoba Hydro in Appendix A.10 describes a particularly successful programme. Brown (2009) provides an overview of on-bill financing programmes undertaken by US utilities. Outside of North America, the Tunisian PROSOL programme for solar water heaters includes an on-bill financing component (see Appendix A.4).

In the UK, a scheme similar to an on-bill loan programme is being suggested as central component of the proposed 'Green Deal': under the proposed scheme called 'Pay-as-you-save', building users would borrow money from private lenders and would pay the loan back through their energy bills. If they moved out and stopped being responsible for the energy bills of the property, the financial obligation would move to the next bill-payer (UK DECC, 2010). The suggested scheme closely resembles the approach taken for the Tunisian PROSOL programme (see Appendix A.4).

SWOT analysis

STRENGTHS

For building owner/user:

- Overcomes the barrier of high up-front costs, as it allows building owners to apply RET and EE measures with limited or no own capital outlay.
- Interest rates may be significantly lower than if the building owner would arrange for financing himself.
- If set up well, the programme can be very simple for the building owner or user.
- Generally, investments are structured in a way that there are net energy cost savings for the building owner.
- For the customer, it is easy to see potential cost savings on his utility bill by comparing bills before and after the installation of RE and EE measures (Brown & Conover, 2009).
- In an on-bill tariff programme, the liability stays with the utility meter, and may thus not be classified as personal debt. As a consequence, lower income borrowers, or persons who are unable to take on additional debt, may also participate in the programme (Brown & Conover, 2009).
- Frequently, the utility guarantees for the performance of the installed equipment; thus the building owner or user is not liable if technical problems occur. In on-bill tariff programmes this is more frequently the case than in on-bill financing programmes (Brown, 2009).

For the utility:

- In on-bill tariff programmes, utilities have the ability to disconnect customers from utility services in case of default on the loans.
- Linking payments to utility bills offers a relatively secure way of recovering the loan. As a consequence, it may be possible to offer attractive interest rates due to the lower default risk. However, engaging with lower income property owners may still increase the risk of default for the utility.
- An on-bill financing programme may be a way to increase customer retention in liberalised markets.
- An on-bill financing programme may allow a utility to meet targets sets under energy saving obligations.

WEAKNESSES

For the building owner:

- In the case of on-bill loans, customers must repay the full loan when they sell the property (Brown & Conover, 2009). As a consequence, customers may be hesitant to take on longer term loans. In on-bill tariff programmes where the obligation stays with the meter, loan terms are generally a bit longer.
- The range of RET and EE measures that can generally be financed by on-bill financing alone is limited to measures which are cost-effective with a repayment term of about 10 years maximum. At current prices for conventional energy and RE and EE measures a number of potential measures are still too expensive.

For the utility:

- Generally, utilities are reluctant to enter into wholesale lending business as it is outside their traditional competencies. They consider it risky to use their own capital to make loans to customers. As a consequence utilities undertaking on-bill financing programmes in the US may require short repayment periods of five years or less, which are too short for most potential RET and EE measures in the residential sector (CalCEF, 2009).
- Changes to the billing system are often difficult and costly to implement.
- On-bill tariff programmes require approval of the regulator for the new tariff structure.

OPPORTUNITIES

- A potential change in mindset of utilities towards considering energy efficiency measures of its customers as opportunities, e.g. due to imposed obligations or for reasons of customer retention, may motivate more utilities to become involved in on-bill financing programmes. A utility can also perceive an on-bill financing programme not as an obligation, but as an opportunity to generate returns.
- Especially on-bill tariff programmes can be structured in a way that they are also attractive to renter-occupied buildings, as the repayment of the loan is linked to the utility meter (CalCEF, 2009).
- Utilities can partner with financial institutions that lend money to private persons and small businesses. This could make on-bill financing more attractive to utilities and leverage the competitive advantages of banks in evaluating and managing consumer credit. The Tunisian PROSOL case study is an example of such a successful collaboration between a utility and financial institutions. Increasing interest from private and institutional investors, e.g. from pension funds, may provide opportunities for utilities to form such partnerships.

THREATS

- In cases where the utility provides the capital for financing an on-bill programme, the utility may be concerned about defaults on loans and therefore be unwilling to engage (Brown & Conover, 2009).
- If third parties provide the capital for financing an on-bill programme, there may be a situation where the responsibility for collecting repayments lies with the utility and the financial liabilities with the third party. This situation is relatively unusual and may prevent utilities and third parties to engage in an on-bill financing programme (Brown & Conover, 2009).
- In the US, there are some pending legal issues, for example around the question if a utility is allowed to disconnect customers from its services if they fail to pay (see Nostrand, 2011). Legal issues around the concept can also be expected in other countries where on-bill financing is applied.

Strengths

- Overcomes barrier of high up-front costs
- Secure repayment through utility bill
- Liability may be transferred if it is linked to the utility meter

Weaknesses

- Limited to measures that are cost-effective over 5–10 years
- Changes to the utility's billing system may be difficult to implement
- On-bill tariff programmes require approval of the regulator

Opportunities

- Change of mindset of utilities towards being more interested in EE
- Partnerships between utilities, banks and installers

Threats

- Concerns about defaults on loans
- Pending legal questions, e.g. on ability to disconnect customers

Figure 4.25 On-bill financing – summary of SWOT analysis

This poses a risk especially for the first programmes in a new market until the legal situation is fully clarified.

Discussion and conclusions

On-bill financing programmes are expected to work best if the involved utility has a strong interest in the programme, e.g. because it considers the programme as an opportunity to grow its business or retain customers. The current structure of energy markets where mostly a large part of a utility's revenues are directly linked to energy sales provides a barrier to strong interest by utilities in on-bill financing programmes.

A review of existing on-bill financing programmes in the US and Canada has shown that in these cases the mechanism worked well for small businesses which required simple, turnkey approaches to improve their energy efficiency and for private owners of residential buildings seeking financing for modest energy efficiency measures (Brown & Conover, 2010).

Governments can use various ways to facilitate on-bill financing programmes, e.g. by:

- Mandating or strongly incentivising utilities to implement such programmes, for example by restructuring a utility's revenue structure in a way that rewards energy savings and installation of RET in buildings.
- Clarifying legal issues around liabilities created through on-bill financing programmes.
- Partnering with utilities in providing access to capital for the programme, e.g. through revolving funds.
- Offering the opportunity to combine the programme with subsidies to enable the installation of a wider range of RET/EE measures.

Utilities planning an on-bill financing programme can consider partnering with specialists such as installers, service companies undertaking energy audits, banks and ESCOs; especially the role of installers and contractors may be critical for successful programmes. A review of two US programmes, for example, suggests that the success of these programmes depended on the establishment of strong relationships with partners (Johnson et al., 2011). The value of partnerships is also demonstrated in the Tunisian PROSOL example which brought together a number of relevant actors active in the specific market segment. In any case, it is advisable for utilities to discuss a planned on-bill financing programme with the energy market regulator, even if it is an on-bill loan rather than an on-bill tariff programme (Brown, 2009).

Box 4.2 Case in point: PROSOL

The Tunisian programme PROSOL creates an innovative structure of support mechanisms to create successful business models around the financing, supply and installation of solar water heaters (SWH). The initial set up of PROSOL included:

*   A capital cost subsidy for 20 per cent or more of the initial cost of the SWH.
*   Reduction of interest rates through an agreement with participating commercial banks to charge lower interest rates.
*   An on-bill financing mechanism where customers who install a SWH repay the loan via their electricity bill over a period of 5 years.

PROSOL has generated significant opportunities for companies selling and installing SWH in the country and has stimulated high growth in the number of installed equipment (see Appendix A.4 for more details).

Leasing of renewable energy equipment

Description

Leasing enables a building owner to use a renewable energy installation without having to buy it. The installation is owned or invested in by another party, usually a financial institution such as a bank. The building owner pays a periodic lease payment to that party (Activum finance, 2011). Leasing therefore resembles renting (Brealey & Meyers, 2003) or hiring of a renewable energy technology.

Generally, the financial institution (or another actor who offers the lease, i.e. the lessor) remains owner of the asset during the lease period. However, several types of leasing are possible, which differ in ownership and other economic, legal and fiscal conditions (Clearsupport, 2008; Bleyl & Schinnerl, 2008b). There are two main types of leases: operational lease and financial lease.

Leasing can be a central component of the business model of an Energy Service Company which has limited own capital and therefore also only limited access to debt, but may lease equipment from a financial institution. The ESCO then installs the equipment at the premises of its customers as part of the services that it offers. However, building owners may also finance RET via leasing without the involvement of an ESCO.

Leasing can also be a central component of the business model of a company that introduces a specific new technology to the market. The company that provides the technology can offer it to property owners via a leasing arrangement, including a service and maintenance package.

Market segments

Leasing could be applicable to all types of buildings.

Applicable technologies

Leasing is common for certain moveable goods such as cars, but also other equipment, like ships and airplanes (Activum finance, 2011; Bleyl & Schinnerl, 2008b). Leasing may be available for energy equipment and installations like condensing boilers (AgentschapNL, 2010), small and micro-CHP systems or solar energy equipment in buildings. The case study in Appendix A.8 describes, for example, leasing of a large heat pump system in an office building. However, overall, leasing is not frequently used for RET. One reason for this is that not all RET can be leased. Generally, any equipment which is an integral part of a building is to be owned by the building owner. If installed technologies become part of the building, an operational lease is impossible because for this type of lease the ownership has to remain with the lessor, i.e. the actor offering the lease. Another reason is that regulation usually requires that after the leasing period, an asset can be re-used in a reasonable state at a different time and place. This criteria is referred to as 'fungibility' (Bleyl & Schinnerl, 2008b). RET systems such as soil- or water-based heat pumps do not meet this criterion as the complete system cannot be removed without substantial damage. Similarly building insulation, which is often a very suitable EE measure, cannot be removed after the end of the lease term.

Actors

Usually an ESCO or a building owner takes a lease while a financial institution or bank provides it (Bleyl & Schinnerl, 2008b). Also, a company aiming to introduce a new technology to the market may offer leasing of this technology to a building owner or user.

Organisational and financial structure

Leasing involves a temporary financing contract between a party (the 'lessor', such a bank) that provides an asset (such as a renewable energy technology) to another party (the 'lessee', such as a building owner) which wants to use the asset for a certain period. The lessee pays in exchange a periodic payment for the lease to the lessor. In this way, the lessees do not need to make an investment and do not take on any debt, although they still need to account for the liability from the expected lease payments.

The choice for leasing or another financing option is case specific and depends on many aspects such as:

- *The direct financing costs* compared to the lease payments
- *Legal aspects*, such as the ownership situation and its implications, and the conditions for contract cancellation
- *Securities required by the lease provider*
- *Tax issues*
- *Accounting issues*, e.g. who activates the investment, etc.
- *Other aspects*, e.g. transaction costs, comprehensive consultancy, etc. (Bleyl & Schinnerl, 2008b).

The following examples illustrate three potential leasing arrangements.

Examples involving a bank and a property owner or ESCO

The first two examples below assume that a bank acts as the lessor of RE equipment. In the first example a building owner leases a solar water heater directly from a bank, which owns the equipment. In exchange the building owner pays a periodic lease rate during the contract period which includes a lease instalment and interest share (Bleyl & Schinnerl, 2008b; Gray & Needles, 1999) (see Figure 4.26)

In the second example, an ESCO undertakes the negotiations with the financial institution, provides additional services to the building owner and remains the lessee of the equipment, which is still owned by the financial institution (Figure 4.27). The advantage of the involvement of an ESCO is that the ESCO can act as a facilitator.

The example of the lease of a large heat pump system described in Appendix A.8 illustrates a slightly different structure where the involved ESCO is the lessor and economic owner of the RET.

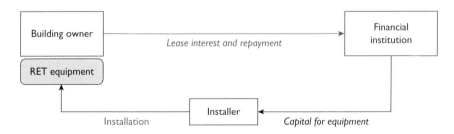

Figure 4.26 Lease agreement without involvement of an ESCO

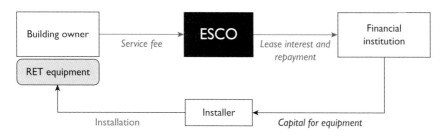

Figure 4.27 Lease agreement with involvement of an ESCO

Example involving a technology provider

In the third example, a provider of a specific technology, e.g. a micro-CHP system, leases the system to private or commercial customers (Figure 4.28). This approach is mostly used by companies who want to bring a new (energy) technology to the market, and have to compete against established technologies, traditional institutional areas of influence and potentially long supply chains to individual customers (Foxona et al., 2005). The technology provider usually also provides operation and maintenance services for the equipment.

Ownership of the assets

In leasing arrangements there is a difference between legal ownership and economic ownership (AgentschapNL, 2010). In the first two examples, the bank always has legal ownership over the asset (at least during the leasing period), meaning that it has ultimate decision power over the asset. This aspect distinguishes a lease from a loan, because with a loan the building owner (or ESCO) holds legal ownership (the money to buy the asset is 'rented', not the asset itself).

Economic ownership over the asset can however be either at the building owner/ESCO or the bank. Economic ownership implies that a party receives economic benefits, such as tax benefits from deducting lease rates, but also bears economic risks from the asset, such as damage to the equipment.[26] The

Figure 4.28 Leasing arrangement between a company distributing a specific technology and a building owner

economic owner is obliged to capitalise the asset on its balance sheet (Bleyl & Suer, 2006).

The ownership situation is a main difference between an operation and financial lease. With an operational lease the bank owns the economic rights, with a financial lease the ESCO/building owner does. Thus only operational leasing is regarded as off-balance sheet financing (IEA, 2010) as the value of the lease contract does not appear on the building owner's balance sheet. With a financial lease, it is possible that after the lease period the lessee also become the full legal owner of the leased asset (Activum finance, 2011). Differences between operational and financial lease are explained in detail in Bleyl and Schinnerl (2008b).

Provision of additional services

Additional services provided are another difference between an operational and financial lease. In the case of an operational lease a building owner is not responsible for e.g. financial accounting and administrative obligations. The bank (or other lessor) performs these as lease services to the building owner (Clearsupport, 2008). With a financial lease, the building owner is responsible for these arrangements. An operational lease thus offers more outsourcing of services and comfort to the building owner.

Differences between a lease and a loan

There are also differences between a lease and a loan:

- Leasing is generally more expensive for the ESCO/building owner than a loan (Clearsupport, 2008),[27] but in return provides more services to the building owner, for example, the lease service package can include technical consultancy. Administrative costs e.g. for negotiating terms are expected to be higher than for a loan as a lease contract for energy equipment is generally more complex than a loan (IEA, 2007). However, the Dutch energy service company Essent (2011), who facilitates leasing contracts for energy equipment between its clients and a financial institution, indicates the additional costs are limited when standard lease agreements and procedures are established and projects are sufficiently large to have relatively low transaction costs.
- Leasing is considered to be more flexible than a loan, as for example shorter depreciation schemes are possible. For a loan, depreciation times are regulated by tax law.[28] Also, a financial lease contract can usually be set for a shorter term than a loan.
- Tax deductions are possible for both operational leases (deduction of lease rates) and financial leases. Leasing can, for example, enable organisations to make use of subsidies or tax advantages for which they would

normally not be eligible, e.g. because their legal status does not allow for it, but the subsidy or tax advantages can be claimed by the lessor and passed on to the lessee (Clearsupport, 2008; AgentschapNL, 2010).

Existing markets and policy context

Country specific regulation on leasing, e.g. regulation on tax implications and depreciation, may have a substantial impact on the decision if leasing is an attractive option for a building owner or ESCO (Clearsupport, 2008).

Overall, leasing is not frequently used to finance RET. In Austria, leasing is for example typically applied for large-scale renewable energy projects, e.g. wind farms but is not common for RE in buildings. In the Netherlands, only some energy service providers are known to rent or lease solar water heater to their clients (Milieucentraal, 2011). Energy companies usually rather rent or sell RET instead of offering leasing arrangements.

Financial institutions are generally found to be not very open to leasing energy technologies (Fina-ret, 2008), probably because energy is not part of their core business, because they are not familiar with the implications of leasing of RET, or because they may not want to take the risks associated with ownership of the technology. According to Essent (2011) the credit-worthiness of the client and a financially sound business case are the main requirements by a bank for leasing. Public organisations would therefore be more easily accepted as clients than commercial actors.

SWOT analysis

STRENGTHS

For a building owner/user:

- Leasing of equipment provides a building owner or occupier the opportunity to use this equipment without initial investments, thus helping to overcome the barrier of high up-front costs.
- By leasing via an energy service contractor, the building owners may profit from additional services such as specific financial, legal, fiscal and administrative consultancy, and operation and maintenance services. This may imply lower financing costs (due to better understanding of the risk by the lessor) and lower transaction costs and effort for lessee (Bleyl & Schinnerl, 2008b).
- Leasing terms are generally more flexible than for a loan. For example, banks are more flexible in setting lease periods.
- Leasing can be structured in a way that makes optimal use of subsidies and tax deductions.
- By leasing off the balance sheet, the building owner or ESCO can access additional financing to invest in other assets (Bleyl & Schinnerl, 2008b).

For an ESCO that leases equipment from a financial institution:

- Leasing is more flexible than debt because it can be used to finance part of a project. With a loan (such as a mortgage) this is more difficult, as ownership rights of parties involved determine the possibilities of mortgage financing (Essent, 2011).

For a technology provider that leases out equipment:

- The company can keep the responsibility for maintenance which may be important for technologies which are just entering the market or for technologies that are very maintenance intensive, e.g. CHP systems based on gas engines.
- Leasing provides an opportunity to distribute a technology that is too costly to be sold but that does generate cost savings over its lifetime.
- Leasing provides an opportunity to distribute an energy technology in which customers may not yet have trust because it is new and considered risky, e.g. fuel cells.

WEAKNESSES

- Financial institutions are hesitant to become active in leasing out RET as energy is not their core business and they may not want to assume operational risk.
- Not all renewable energy projects or technologies qualify for leasing, as it must be possible to remove the technology from the building.
- A bank might put requirements on the building owner, for instance a maintenance requirement, or a provision of extra collateral in addition to expected project cash flows (Bleyl & Schinnerl, 2008b).
- Although leasing periods can be set flexibly, leasing is typically undertaken for a fixed leasing period during which termination of the lease is not possible (Clearsupport, 2008).

OPPORTUNITIES AND THREATS

- Changes in country specific regulations, such as accounting and tax regulations, may impact the attractiveness of leasing.

Discussion and conclusions

In general, leasing is not commonly used for renewable energy projects for reasons of suitability of RET for leasing and due to its lacking attractiveness to market parties.

For a building owner, the main advantage of leasing is that he can use the leased equipment without having to invest in it. Although not part of a leasing

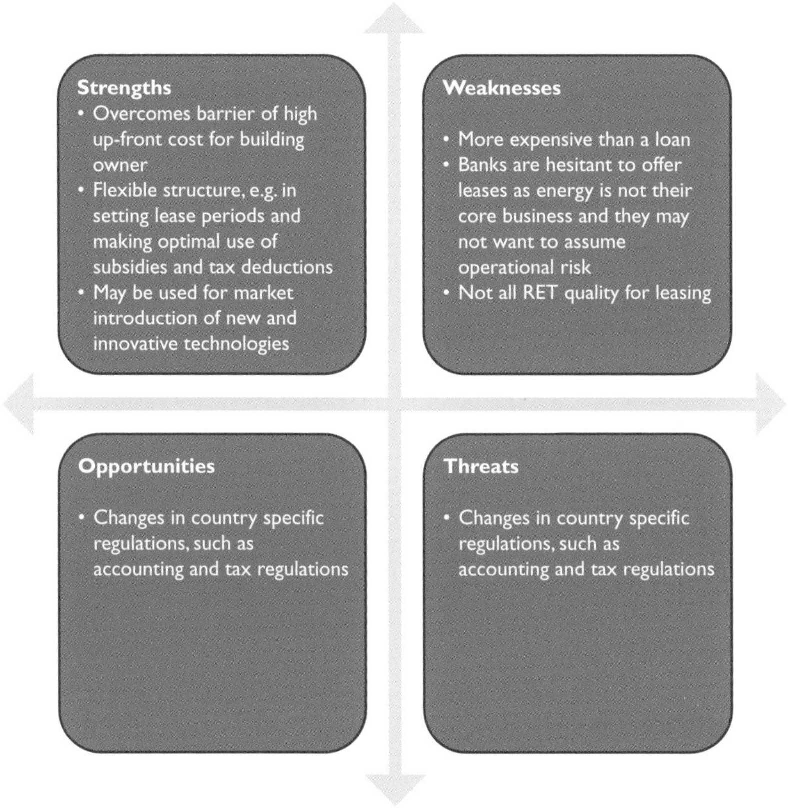

Strengths
- Overcomes barrier of high up-front cost for building owner
- Flexible structure, e.g. in setting lease periods and making optimal use of subsidies and tax deductions
- May be used for market introduction of new and innovative technologies

Weaknesses
- More expensive than a loan
- Banks are hesitant to offer leases as energy is not their core business and they may not want to assume operational risk
- Not all RET quality for leasing

Opportunities
- Changes in country specific regulations, such as accounting and tax regulations

Threats
- Changes in country specific regulations, such as accounting and tax regulations

Figure 4.29 Leasing of RET – summary of the SWOT analysis

arrangement itself, a maintenance contract is frequently offered in combination with leasing. This reduces the effort required by the building owner, but also the technical risk for the lessor. Leasing is generally more expensive than taking a loan or financing the equipment otherwise. It is, however, difficult to assess the exact comparative advantages or disadvantages of a lease over other financing options on a general level. This requires a detailed assessment of several financial aspects, including tax and accounting regulations.

Specifically for ESCOs which often have limited own capital and therefore also limited access to debt, leasing of equipment offers opportunities to offer more comprehensive services.

At the moment there is no clear evidence that leasing contributes significantly to an increased use of RET in buildings. Leasing may offer opportunities for the introduction of new technologies to the market. Leasing is for example used for the distribution of small and micro-CHP systems (see Box 4.3 and Appendix A.7). As demonstrated in the example in Appendix

A.7 by the cooperation between Volkswagen and LichtBlick innovative business models are emerging that make use of leasing out of equipment or similar arrangements. Government support for such new and innovative business models has to be targeted to the specific technology and market environment. Aggregation of small CHP systems in order to manage imbalances in electricity supply and demand could for example be supported through an improved IT infrastructure such as smart grids and smart meters.

Box 4.3 Case in point: Market introduction of small and micro-CHP systems

Leasing is being applied by companies in different countries and contexts for the market introduction of small and micro-CHP systems. Appendix A.7 illustrates two case studies from Japan and Germany.

Business models based on energy saving obligations

Description

Energy saving obligation schemes, sometimes also referred to as 'white certificates', are not a business model in itself, but a policy instrument that obliges energy companies to realise energy savings. It stimulates business models based on financial incentives offered by energy suppliers to building owners, renters or to energy service companies. The obligations mandate energy suppliers to realise energy savings at the level of end-users. Though primarily aimed at energy savings and energy efficiency, this regulation could potentially also be an important driver for investments in RET in the built environment.

Energy saving obligations may stimulate energy companies to develop business models to realise the mandated energy savings, e.g. by offering new energy efficiency services to customers (often in partnership with electricians and installers) as is the case in France, or by 'outsourcing' a large part of the obligations to energy service companies and thus creating a significant ESCO market, as is the case in Italy (Boot, 2009; ECN, 2011b).

Mostly, energy companies are allowed to pass on the costs of the EE measures (and potentially RET) to all consumers via higher energy prices. Thus, the saving obligations can also be considered a financing mechanism for the development of energy savings or RET.

Market segments

Energy saving obligations mostly target existing buildings. Within this segment, the specific sub-segment targeted differs by country. In the UK, the savings must, for example, be realised in residential buildings. Forty per cent of savings need to be realised in low income households to reduce energy poverty (Ofgem, 2010). (A more detailed description of the UK energy supplier obligation is given in Appendix A.9.) The French regulation also mostly targets the residential sector (90 per cent of realised savings in the first obligation period) and includes fuel poverty as a consideration (JRC, 2011). In Denmark the residential sector is not the primary target for savings but rather other end-use sectors such as industry.

Applicable technologies

Most existing schemes focus on energy efficiency measures. However, energy saving obligations could be extended to cover RET. In the UK scheme, insulation is for example the dominant measure, followed by energy efficient lighting. The share of RET is small (Ofgem, 2010; see also Appendix A.9). Until March 2011, 2,000 m² solar water heating and 5,500 heat pumps have been installed (Ofgem, 2011).

In France most savings are realised by installing efficient condensing boilers. A little more than 10 per cent of the certificates were generated through the installation of heat pumps (JRC, 2011).

Requirements on eligibility criteria can restrict the deployment of RET. In Denmark and Flanders, for example, only the volume of savings realised in the first year after the measure is counted towards the target. This stimulates measures with short payback times, whereas RET usually have longer payback times.

Actors

Directly involved actors are energy companies, usually energy suppliers or distributors on whom the obligation is imposed. In the British and French schemes the obliged parties are energy suppliers; in Denmark, Flanders and Italy the obliged parties are network companies. Frequently, the utilities enter into partnerships with service providers such as ESCOs or installers. In the UK, energy suppliers either offer energy services themselves, or hire external companies (JRC, 2011). In Italy and Denmark, obliged energy companies are not allowed to implement energy saving projects themselves (JRC, 2011). The measures are implemented at the clients of the energy companies, or the end-user sectors. Government is involved in setting the regulatory framework.

Organisational and financial structure

The business models emerging from energy saving obligations vary considerably in design. The basic organisational structure is shown in Figure 4.30.

Energy companies, or the partners to whom the energy company outsources the obligation, need to create (financial) incentives for consumers to voluntarily implement energy savings at their premises. In Italy, France and the UK, subsidies for saving measures are the most common financial incentive (Bertoldi et al., 2009). In order to be able to finance the incentives offered, energy companies are generally allowed to charge higher energy prices to all customers. In many cases, financial incentives for energy consumers are combined with information measures.

There are a number of different design choices for energy saving obligations schemes. Examples are whether obligations involve trading and white certificates (official documents certifying that a certain reduction in energy consumption has been achieved), which actors are obliged parties, what kind of target (e.g. primary or final energy) is set and for which period, what eligible savings are, how flexible the scheme is with regards to eligible savings, and how monitoring and control of the obligations scheme is managed. There is no consensus, yet, on the best design in a specific situation.

In the UK, energy suppliers are estimate to have invested 3.2 billion pounds between 2008 and 2011 in saving measures at households. Many saving measures are delivered by the suppliers at no cost to the home owner (Ofgem, 2010; see also see also Appendix A.9). Suppliers in the UK use several ways to persuade home owners to take saving measures. The most important is subsidising the installation of measures by installers. Other suppliers are marketing programmes to private home owners, enter into partnerships with

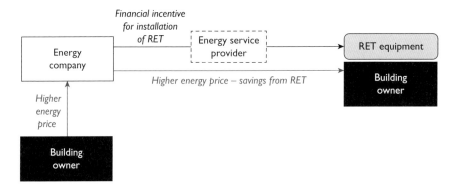

Figure 4.30 Schematic depiction of energy saving obligations, depicting one building owner who profits from the obligation by lower energy prices after an installation of RET/EE measure and another building owner who only pays a higher price for his electricity

local public organisations (social housing corporations) or with other government saving programmes (Ofgem, 2010).

In France, consumers do not directly pay the costs of saving measures. Energy suppliers provide financial incentives to customers such as subsidies and soft loans (JRC, 2011). Investment can be earned back via energy bills, but since energy tariffs are regulated the regulator needs to take these investments into account (Bertoldi et al., 2009). Most energy suppliers have set up their own programmes.

Existing markets and policy context

As of 2011 four EU Member States and one region in the EU have an energy saving obligation scheme in place: the United Kingdom, Denmark, France, Italy and the Flemish region in Belgium. Some countries seem to be planning or considering energy saving obligations schemes, e.g. Poland, Ireland, Bulgaria, Romania, Germany and Portugal. Research in Sweden advised against the implementation of an obligation scheme following an ex ante evaluation (ECN, 2011b).

Energy saving obligations also exist in the US, as of 2009 in 15 states. Some of these are combined with renewable portfolio standards (RPSs), which mandated energy suppliers to produce renewable electricity, mostly through large-scale wind, solar, biomass and geothermal installations (US EPA, 2011a). As of 2011, there are RPSs in 43 states (Dsireusa, 2011). However, it is not expected that a significant amount of RET is deployed in buildings under these schemes. Although small-scale renewable energy production is possible, it is not widely eligible under the existing RSPs. Moreover, the financial incentives offered to building owners are mostly not attractive. Concrete data on the deployment of small-scale RET in buildings under RPSs does not exist (US EPA, 2011b).

SWOT analysis

STRENGTHS

For governments:

- Energy saving obligations are a means for government to catalyse energy savings without having to use government funds to finance and administer the programme. Instead, costs are paid for via higher electricity prices.
- Energy saving obligations mandate energy suppliers to take action thus addressing the barrier 'lacking intrinsic interest by energy companies', which implies that due to their revenue structure, energy suppliers have no interest in efficiency measures.

- Energy companies are generally in a strong position to implement the obligations as they already have direct contacts to households, which are their customers, and are often in a healthy financial position.

For building owners:

- Energy saving obligations reduce the barrier of high up-front costs or cost effectiveness of EE measures or RET for the building owner, as obliged energy companies provide financial incentives.
- The building owner is usually supported by energy companies with information and advice, and organisation of the installation.

For ESCOs:

- Energy saving obligations may create demand for energy efficiency services.

WEAKNESSES

For building owners and the public:

- Energy saving obligations lead to higher energy prices as energy companies need to recover their costs. However, the price increase may be limited (Bertoldi & Rezessy, 2009). At the same time the level of ambition of the obligation is linked to the level of energy price increases.
- There is a problem of equity, as not all consumers profit from the savings, but all pay higher energy prices.

For governments:

- Policy implementation can be costly depending on the design of the verification approach and depending if and how trading is allowed within the obligation scheme.
- Energy saving obligations mainly catalyse EE measures which generally have lower up-front costs or are more cost effective than RET, unless the latter is supported by additional incentives.
- Energy companies are likely to first target technologies or even building segments with the lowest improvement costs (realise the most cost-effective measures). The majority of the implementation costs may thus be postponed to the future, which can overstate the scheme's cost effectiveness in the beginning and can make the realisation of saving targets in the future more difficult.

OPPORTUNITIES

- Countries have the opportunity to design the obligation scheme in a way that also stimulates RET in buildings, e.g. by incentivising RET, for example awarding it with more certificates, or by only allowing RET instead of EE measures in specific situations.
- Energy saving obligations may strengthen the competitive position of energy companies as the obligations may incentivise companies to increase their scope of services and focus on maintaining long-term customer relationships. However, this requires energy companies to make an effort to adapt their business model (Bertoldi & Rezessy, 2009).
- The proposed EU Energy Efficiency Directive, replacing the current EU Energy Services Directive, is expected to introduce mandatory national energy saving obligation schemes in every EU member state.
- Increasing energy market liberalisation in some countries can enable a fair cost recovery from consumers by energy companies as energy price increases need to be competitive (i.e. not too high).

THREATS

- Utilities may have little intrinsic interest in implementing the savings obligations as it is not part of their traditional core business and may negatively affect their revenues as energy sales go down. This may lead to programmes that do not achieve the full savings potential or that are not designed in a cost-efficient way.
- Governments may be hesitant to approve increases in energy prices.

Discussion and conclusions

Energy saving obligations are not a business model by themselves, but a policy instrument that may lead to business models based on financial incentives offered by energy companies to building owners, building users or to energy service companies. Energy saving obligations can be a powerful instrument as they force energy companies, i.e. energy suppliers or distributors to actively pursue EE measures and potentially RET. The advantage for governments is that they can 'outsource' the financing and implementation of energy efficiency measures and only need to set the regulatory framework and undertake the monitoring of the implementation of the directive.

There are many approaches in designing energy savings obligations, e.g. with or without associated trading of certificates, targeting different (sub-) sectors and technologies, and involving different actors for implementation. Given the limited experience with the scheme to date, it is too early to tell which approaches are most effective and efficient under which background conditions.

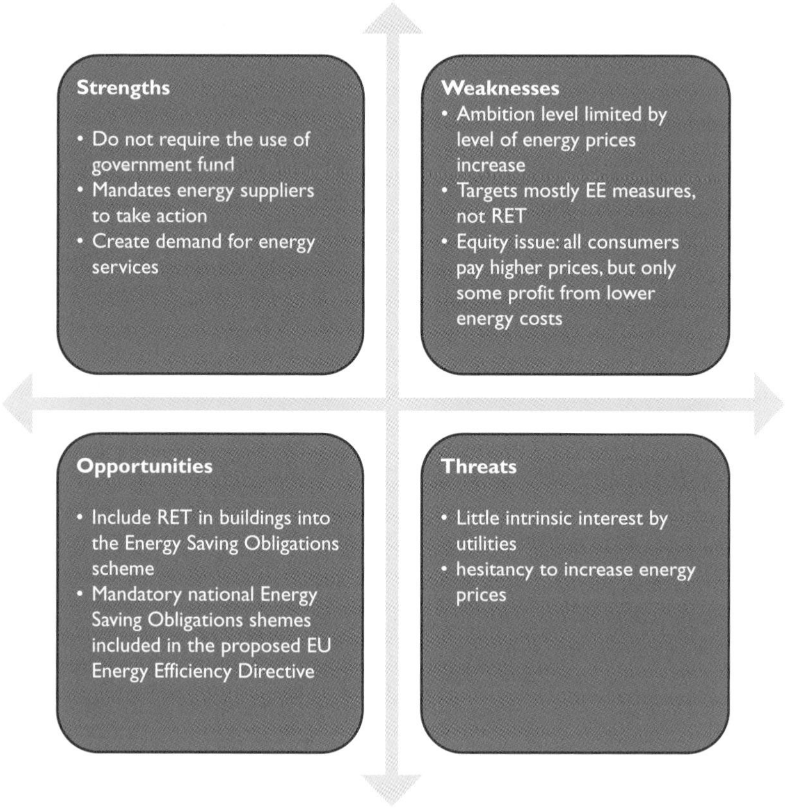

Strengths

- Do not require the use of government fund
- Mandates energy suppliers to take action
- Create demand for energy services

Weaknesses

- Ambition level limited by level of energy prices increase
- Targets mostly EE measures, not RET
- Equity issue: all consumers pay higher prices, but only some profit from lower energy costs

Opportunities

- Include RET in buildings into the Energy Saving Obligations scheme
- Mandatory national Energy Saving Obligations shemes included in the proposed EU Energy Efficiency Directive

Threats

- Little intrinsic interest by utilities
- hesitancy to increase energy prices

Figure 4.31 Energy saving obligations – summary of the SWOT analysis

In the EU, energy saving obligations are expected to become widely used if the proposed EU Energy Efficiency Directive is implemented.

Notes

1 See for example EC (2006), Bertoldi et al. (2007), EN (2009), Satchwell et al. (2010).
2 More details on financing options for energy contracting projects can be found in Bleyl and Schinnerl (2008b), who introduce the customer demand profile methodology. This is a tool to define and structure financing needs from the customer perspective and can be used as a checklist to compare different financing options.
3 See www.grazer-ea.at
4 See www.berliner-e-agentur.de
5 Deemed savings are an approach to estimating energy and demand savings, usually used with programmes targeting simpler efficiency measures with well-known and consistent performance characteristics. This method involves multiplying the

number of installed measures by an estimated (or deemed) savings per measure, which is derived from historical evaluations. Deemed savings approaches may be complemented by on-site inspections (source: http://www.epa.gov/statelocal climate/resources/glossary.html).

6 Thus energy contracting can be considered a tool for moving towards a thinking of 'economics of stock'. 'Economics of stock' imply major long-term investments in facilities to produce and deliver fuels and electricity, with concomitant long-term finances and contracts, business relationships and risks (Patterson, 2010).

7 Note that there are also energy suppliers that offer some energy services (see e.g. example in Box 5.1) but without the performance guarantee offered by an ESCO. These situations are not depicted in the graph.

8 Note that this business model analysis is based on the use of voluntary building certification schemes. It does not include mandatory energy performance certification as required by the EU Directive on the Energy Performance of Buildings and translated into national law in the EU member states.

9 The terms 'green' and 'sustainable' are often used interchangeably in this context (Nelson et al., 2010). In this report, only the term 'green' is used.

10 See http://www.breeam.org/page.jsp?id=301

11 See http://www.gbci.org/main-nav/building-certification/resources/fees/current.aspx

12 See http://www.rednews.com/index.php/2011/01/one-billion-square-feet-of-leed-commercial-buildings-2/

13 See http://www.breeam.org/page.jsp?id=66

14 See http://oee.nrcan.gc.ca/residential/personal/new-homes/r-2000/about-r-2000.cfm?attr=4

15 See http://oee.nrcan.gc.ca/residential/business/new-homes/new-homes-initiative.cfm?attr=0

16 See e.g. BREAAM and LEEED case studies at http://www.breeam.org/case-studies.jsp and http://www.usgbc.org/DisplayPage.aspx?CMSPageID=1721. The Rodney Phase 1 residential development (http://www.breeam.org/page.jsp?id=324) carries for example a BREAAM excellent rating without integration of low or zero carbon technologies.

17 See http://www.bca.gov.sg/greenmark/green_mark_buildings.html

18 See http://www.breeam.org/

19 In the EU, publicly owned or occupied buildings represent about 12 per cent of all building area (Ecorys et al., 2010).

20 CFV estimated in 2010 that housing corporations can acquire capital via CFV, at 1.5% lower financing costs than on the market (ECN, 2011a).

21 In the Netherlands a more pragmatic assessment is allowed if no energy label is available. In such cases the age of the dwelling determines the amount of 'points' awarded (Eerste Kamer, 2011).

22 Assessments are comparable to loans as the property owner pays off its debt in instalments over a period of various years. But legally, PACE assessments are not considered to be loans (NREL, 2010).

23 In the US, property tax payments are made annually or in arrears. Payment modalities may be different in other countries.

24 A lien is a legal claim against an asset to secure a loan.

25 For further information, see for example http://www.institutebe.com/clean-energy-finance/melbourne-environmental-upgrade-agreement.aspx

26 In other words, economic ownership of an asset defines who bears the risk from an increase or decrease of value of an asset.

27 Examples of reasons are higher asset risk (responsibility of operation and maintenance remains at the lessee), flexibility of a leasing contract (e.g. banks are more

flexible in setting the lease period) and additional leasing services that are usually provided (e.g. subsidy acquisition).
28 In practice, depreciation periods are usually fixed for 15 years depending on the nature of the asset and expected lifespan.

Chapter 5

Synthesis: business models, barriers, market segments and actors

This chapter summarises and evaluates key features of the business models discussed in the previous chapter and puts them into a larger perspective. The analysis looks at which barriers are addressed by the business models described in this book, in which market segments the business models can work and which actors are directly involved. In this way, it addresses the question how the analysed business models can stimulate an increased deployment of RET in the built environment. The chapter ends with conclusions and an outlook.

Which barriers are addressed by business models?

This section discusses which barriers are addressed by the business models analysed in this book (see also Table 5.1). For each barrier it describes whether the barrier can be addressed via one or several of the business models and how. Table 5.1 illustrates which barriers the business models described here are able to address. Generally, some barriers such as 'high up-front costs', 'higher risk from RET than from conventional technologies' and the 'hassle factor' are addressed by outsourcing these issues to a third party who is better equipped to deal with them than a building owner. Other barriers cannot be addressed by business models. Regulatory barriers, for instance, can only be removed by policy makers, and 'low priority of energy issues' is mostly related to the mindset of decision makers. Furthermore, there are barriers, such as high transaction costs, that could be addressed by business models, but not by those described here (see below).

The individual barriers can be addressed as follows.

Low or no return on investment

Generally, most of the business models analysed here cannot change payback periods of EE measures or RET. Thus, on a large scale, these business models work only if RET are cost-competitive over the assumed life-cycle of the equipment compared to traditional energy sources. There are some exceptions to this rule such as a deployment of RET/EE measures due to improved corporate reputation from 'green' buildings. However, this exception tends to be limited to niche markets and is insufficient for large-scale deployment. Other exceptions are situations where a business model directly depends on a financial incentive, and business models that create economies of scale, lowering the costs for RET. Note that economies of scale are not an important element in the business models analysed in this study.

Thus, for most business models to be successful, the barrier 'low (or no) return on investment' needs to be overcome first. Today, many cost-effective opportunities for a deployment of RET and EE measures (see e.g. examples in Box 5.2) exist already, i.e. in these cases the barrier low (or no) return on investment is overcome. However, other RET are not yet competitive and depend on financial or policy support. Without such additional support most business models[1] only lead to a deployment of cost-effective technologies and the business models address other barriers than 'low or no return on investment' which prevent a larger deployment of RET/EE measures.

The business models based on new revenue schemes, i.e. 'Making use of a feed-in remuneration scheme', 'Developing properties certified with a 'green' building label' and 'Profiting from rent increases after the implementation of EE measures', address the barrier of low return on investments by compensating for the low returns via additional revenues. Alternatively, policy makers may deploy a variety of other financial incentives such as subsidies or soft loans. RET can also be mandated via an obligation. Examples for such obligations are the Spanish solar thermal ordinance and the Renewable Portfolio Standards in the US and Japan. Another approach to overcoming the barrier of 'low return on investment (ROI)' could be to outsource the investment to a third party who is able to improve the ROI through a lower cost structure.

High up-front costs and lacking access to capital

The barriers of 'high up-front cost' and 'lacking access to capital' can be addressed through outsourcing the investment to a third party, which recuperates the investment plus capital cost over a project term oriented at the lifetime of the equipment. These barriers are most frequently addressed by the business models analysed here (see Table 5.1), i.e. by the product service system/energy contracting business models and by the business models based on new financing schemes (PACE financing, on-bill financing and leasing). The ESCO models and new financing schemes spread the investment costs for RET over the lifetime of the project. This 'life-cycle approach' provides opportunities to building owners with limited investment budgets. However, access to capital comes at a price: The calculations in Box 5.2 demonstrate the large impact of interest rates on the cost effectiveness of investments in RET.

Table 5.1 Barriers addressed by the business models

	Product Service Systems			Revenue models			Financing schemes			
	Energy Supply Contracting	Energy Performance Contracting	Integrated Energy Contracting	Feed-in remuneration	Sale of certified buildings	Profits from rent increase	PACE financing	On-bill financing	Leasing	Energy savings obligations
Market and social barriers										
Price distortion				●	●					
No interest by energy companies									●	●
Low priority of energy issues										
The 'hassle factor'	●	●	●				●	●	●	
Split incentives [1]						●				
Fragmentation building chain										
Information barriers										
Lack of awareness										
Limited information on financing	●	●	●				●	●		
Limited competence of installers	●	●	●							
Regulatory barriers										
Restrictive procurement rules										
Cumbersome building permitting process										
Economic and Financial barriers										
Low or no return on investment				●	●	●				

Table 5.1 continued

	Product Service Systems			Revenue models			Financing schemes			
	Energy Supply Contracting	Energy Performance Contracting	Integrated Energy Contracting	Feed-in remuneration	Sale of certified buildings	Profits from rent increase	PACE financing	On-bill financing	Leasing	Energy savings obligations
High up-front costs	●	●	●				●	●	●	●
Difficult access to capital	●	●	●				●	●		●
Higher risk of RET	●	●	●						●	
High transaction costs										
Incomplete mortgage assessment										

(1) Limited to split incentives between landlords and tenants in the rental market

Higher risk (technical, economical, etc.)

Business models allow decision makers to outsource technical and economic risks of RET implementation and operation to a third party. In the energy contracting models, the ESCO is paid for the output delivered, i.e. the energy supply and/or savings, and takes on the technical and economic risks. As a consequence, ESCOs generally apply proven technologies with which they have sufficient experience. However, the technology applied could be new to the building owner, who may perceive them as high risk. Usually, an ESCO is unable to take on the high risks involved in the introduction of technologies which are under development or being piloted. However, there are examples of companies that introduce new high risk technologies to the market by offering leasing of the technologies in combination with a maintenance agreement (see case study on small and micro-CHP in Appendix A.7).

The 'hassle factor'

The 'hassle factor' may be overcome by outsourcing a (complete) service package to a third party, which allows building owners to avoid having to manage several different actors and interfaces. The energy contracting models are the most obvious example for this as they offer a whole set of services to building owners, e.g. planning, installation, permitting, financing, operation and maintenance. The business models 'PACE financing' and 'on-bill financing' address this barrier to a lesser extent, generally offering a combination of technical advice, access to capital and installation services to building owners.

Lack of information

The involvement of specialised companies addresses this barrier, as specialists typically have a better knowledge of RET and of available financing options and subsidy schemes than building owners, as well as trained resources to implement RET projects.

Split incentives

Split incentives can be overcome through binding agreements between involved actors, in which the parties who profit from RET/EE measures contribute to the required investment costs. ESCO models can address the split incentive barrier, if there is such an agreement between involved parties and if the prevailing legislation allows for this. Split incentives may also be overcome by allowing building owners to recover the costs for investments into RET/EE measures from their tenants. This business model is the only one which specifically targets the barrier of split incentives between landlords and tenants. The feasibility of the model depends on the country-specific regulation regulating the rental sector.

Another split incentive occurs when buildings frequently change ownership, which lowers the incentive for current owners to invest as they may not be able to recover their investment when selling the house. This specific barrier is addressed by the business model 'PACE financing' and partly by the model 'on-bill financing' where the debts stay with the property, specifically the utility meter, when a property is sold.

'Low priority of energy issues' and 'lack of awareness'

Building owners have a central role in all of the analysed business models as they are usually the party who takes the final decision to install RET/EE measures (see Table 5.3). However, no business model addresses the barriers of 'low priority of energy issues' and 'lack of awareness', which keep building owners from taking action. In the EU ESCO market for example, the lack of sufficient projects was assumed to be the main bottleneck for market growth in the past rather than the availability of financing. In other words, not enough building owners had an interest to involve an ESCO. Government may address this barrier by voluntary measures, e.g. supporting project development by taking on the role as a market facilitator for energy services.

However, the business model analysis has found no evidence that the business models alone will drive a significant transformation of the market, maybe except for energy saving obligations, which mandate energy savings from energy suppliers, but are not a business model in itself. Overall it seems that voluntary measures alone are not sufficient to drive a significantly increased deployment of RET in the built environment. Similarly, the currently discussed revision of the EU Energy Services Directive (ESD) reflects that the existing regulatory framework and incentive programmes are not sufficient to reach the EU's 2020 energy savings goals.

Combining business models

The discussion of the barriers that the business models address demonstrates that business models mostly address only a few barriers. This points to the potential for combining business models in order to overcome several barriers and make the models more viable. As demonstrated by the case studies described in the book, many real world examples of business models supporting an increased deployment of RET and EE measures in the built environment indeed combine two or several of the 'single business models' analysed in the previous chapter. Examples include leasing as a financing option in an energy contracting project (see Appendix A.8), or specific financial incentives to support the realisation of energy savings in energy saving obligations.

In which market segments can the business models be applied?

The following section discusses in which market segments the business models analysed in this book work best (see also Table 5.2). The market segments themselves are categorised by specific characteristics and the relevant barriers (see Chapter 2 and Table 2.1). Table 5.2 shows that the analysed business models are diverse in the sense that for each market segment at least two of the models are applicable.

Existing free-standing residential buildings

Existing free-standing owner-occupied residential buildings are characterised by the fact that building owners have the choice to take investment decisions for RET/EE measures, but may not have a high level of interest, knowledge and available capital. The business models based on a 'feed-in remuneration scheme', 'PACE financing', 'on-bill financing', 'leasing (of small-scale technologies)' and 'energy saving obligations' may lead to an increased deployment of

RET/EE measures in this segment, as they address the barrier of 'high up-front costs' and present a business case to building owners through which they can profit from resulting cost savings or additional revenues.

Existing owner-occupied multi-family buildings

In owner-occupied multi-family buildings, space outside of the apartments such as roof and garden space, as well as the outer shell of the building are frequently commonly owned. This limits the options for apartment owners to install RET/EE measures in existing buildings unless a common agreement among the different owners is reached. However, given agreement among owners and assuming a sufficient size (> 20k EUR energy costs/yr or roughly about 10–15 apartments (Eikmeier et al., 2009), energy supply contracting might work in this market segment.

Table 5.2 Market segments in which the business models work

				Product Service Systems			Based on new revenue models			Based on new financing schemes			
				Energy Supply Contracting	Energy Performance Contracting	Integrated Energy Contracting	Feed-in remuneration	Sale of certified buildings	Profits from rent increase	PACE financing	On-bill financing	Leasing	Energy savings obligations
Residential buildings	New	Built by a project developer		•			•	•				•	
	New	Built by the building owner		•		•	•					•	
	Existing	Owner-occupied	Multi-family									•	•
	Existing	Owner-occupied	Free-standing				•			•	•	•	•
	Existing	Rented	Multi-family	•		•	•		•			•	•
	Existing	Rented	Free-standing						•			•	•
Commercial buildings	New	Built by a project developer		•		•	•	•				•	
	New	Built by the building owner		•		•	•					•	
	Existing	Owner-occupied		•	•	•	•			•	•	•	•
	Existing	Rented		•	•	•	•					•	•

Existing rented multi-family residential buildings

Addressing the market segment of rented residential buildings is challenging due to the barrier of split incentives. The business model through which building owners profit from rent increases after renovation is one of the few models that works in this market segment. In cases where the heat for a large multi-family building or a group of buildings is centrally supplied, energy supply contracting may be a viable business model, if the building(s) have a sufficiently large energy demand (>20 k€ annually), as it allows the building owner to outsource energy related services. The business model based on feed-in remuneration schemes may also work as the building owner can take advantage of outside areas such as the roof.

Existing commercial and public buildings

Commercial and public buildings are generally larger than residential houses, thus allowing for larger project sizes. Therefore the energy contracting models may work here, as the associated high transaction costs can be compensated by the size of the contract. Moreover, commercial and public building owners tend to have a more 'professional approach' to energy management than private owners of residential houses. Therefore, commercial and public building owners recognise the added value of the energy contracting models more easily. Public buildings which may use energy contracting models include government offices, hospitals, schools, prisons and sports facilities.

As in existing owner-occupied residential buildings, the business models based on new financing schemes also work in existing owner-occupied commercial buildings. For the models 'PACE financing', 'on-bill financing' and 'business models based on energy saving obligations', the institutions running the initiative (or the regulator who has initiated the programme) decide which market segments to target. The Environmental Upgrade Charge in Melbourne (see Appendix A.3), for example, targets commercial properties, whereas the Berkely First programme (see Appendix A.2) addressed residential buildings.

New buildings

New buildings offer much more flexibility for the integration of comprehensive EE measures and RET into building design than the existing building stock. Moreover, in most OECD countries, new buildings are subject to relatively strict building codes. As a consequence, new building stock tends to be more energy efficient, and currently business models appear to play a smaller role. Energy saving obligations, for example, mostly do not include measures in new buildings. Furthermore, building codes are increasingly tightened and it is expected that there will also be an increasing deployment of RET in new buildings. Business models may support property developers and building owners in fulfilling these obligations.

The energy contracting models are suitable for new buildings (although EPC is rather applied to the existing building stock). Also, feed-in tariffs supporting the deployment of RET are frequently applied in new buildings. The high flexibility in building design of new buildings to make optimal use of available energy and other environmental improvements may be one of the reasons why certification of properties with a 'green' building label has originally focused on new developments.

Who are the actors involved in the business models?

The following discusses which actors are involved in the business models analysed in this book (see also Table 5.3). In doing so, only actors directly involved in the specific business models are considered. Other actors who are involved on a more general level are not covered, e.g. financial institutions who are involved for the installation of most RET in the built environment as they provide the required capital. Only in the business model 'leasing of RET' do financial institutions directly participate in the business model as lessor of the equipment.

Building owners

Building owners have a central role in all of the analysed business models as they are usually the party who takes the final decision to install RET or EE measures (see Table 5.3). Thus, they are directly involved in all of the analysed business models. However, no business model can substitute building owners' strategic decisions to tap into RE and EE resources and to decide for long-term investments into RET or EE measures, either on their own or through outsourcing to a third party.

Government

Government may be involved in business models mainly (1) as regulator, as is the case for the models based on rent increases, feed-in remuneration schemes, energy savings obligation and PACE financing, (2) by providing financial incentives, as may be the case for feed-in remuneration schemes, or (3) as an actor who implements a business model, as is the case with local governments who set up and administer a PACE financing programme. The following models directly depend on a policy intervention: 'making use of a feed-in remuneration scheme', 'building owner profiting from rent increase after implementation of energy efficiency measures', 'PACE financing' and 'business models based on energy savings obligations'. There are also 'market-driven' models in the sense that they do not directly depend on a policy or financial incentive scheme in order to create a business case

Table 5.3 Actors directly involved in the various business models

Actors	Product Service Systems			Revenue models			Financing schemes			
	Energy Supply Contracting	Energy Performance Contracting	Integrated Energy Contracting	Feed-in renumeration	Sale of certified buildings	Profits from rent increases	PACE financing	On-bill financing	Leasing	Energy savings obligations
Building owner	●	●	●	●	●	●	●	●	●	●
Utility								●		●
Tenant						●				
Government				●	●	●				●
Financial institution									●	
Installers								●	●	
Propery developer					●					
ESCO	●	●	●							
Others						●				●

Note: Only actors that are directly involved in the business model itself are marked with a black dot Financial Institutions are e.g. frequently involved for any installation of RET in the built environment. However, only in the business model 'leasing of RET' do they directly participate in the business model as lessor of the equipment.

for a building owner. Examples for such market-driven business models are ESCOs.

Other actors

In Energy contracting business models the two directly involved actors are the building owners and the ESCO, whereas in the business models based on new financing schemes a variety of other actors are active. The new financing schemes differ with regards to the involved actors, but are similar from the point of view of the building owner who takes the decision for installing RET, as these models spread the up-front costs of the RET over the lifetime of the equipment: leasing is initiated by a financial institution like a bank. On-bill financing is initiated by a utility. PACE financing is initiated by the local government. It is not possible to make a general statement about which model works best, as this highly depends on the local circumstances.

Conclusion and further considerations

Conclusions

This study demonstrates that business models can play an important role for increasing the deployment of RET in the built environment. They provide opportunities for building owners (e.g. financing of up-front costs, outsourcing of risks) and in many cases require only a supporting role by government, e.g. through changes of legislation such as allowing for rent increases after EE measures. However, business models alone will not lead to a significantly increased deployment of RET. They generally only lead to a deployment of cost-effective technologies because most business models discussed here cannot improve the returns on investment of RET and EE measures by themselves, but only work if there is a favourable business case. Moreover, business models cannot address all barriers, e.g. no business model addresses the barrier of 'low priority of energy issues', which keeps building owners from taking action.[2] Thus a strong role of policy makers is still required.

The analysis also shows that the built environment is a complex sector in which some barriers for increasing the deployment of RET differ among market segments. Earlier in this chapter we summarised and discussed which barriers are addressed by the new and innovative business models analysed in this book, and in which market segments the business models work. The following gives a short overview of the main conclusions from this discussion:

- The barriers of high up-front costs and access to capital can be tackled with ESCO models or leasing for existing and new large commercial, residential and public buildings and with PACE or on-bill financing for small residential and commercial buildings. These business models make a life-cycle approach possible which from the point of view of the building owner spreads the investment costs over the project lifetime. The investment costs are compensated by the benefits of energy savings.
- For business models to work in rental buildings, the split incentives barrier must be solved. This book describes changes in legislation allowing building owners to increase rents for energy efficiency improvements, which can be applied in jurisdictions where the rental sector is regulated, and specifically in the social housing sector where rent levels are determined according to a set of criteria. To soften the social effects of the

measure, for the tenants the benefits of energy savings should be higher than the rent increase.
- ESCO models, leasing, PACE or on-bill financing and rent increases only work for RET or EE measures that are cost effective. Today, there are already many cost-effective opportunities for a deployment of RET and EE measures (see e.g. Box 5.2). For technologies where this is not (yet) the case, business cases may be based on supporting policy measures such as feed-in remuneration schemes (see Chapter 4). Moreover, 'green' certification of buildings can stimulate investments in RET also when they are not cost-effective. But because such certification is voluntary, it is only expected to work in niche markets.
- Energy saving obligations are introduced by governments to stimulate EE measures, energy services and the participation of energy suppliers. This policy measure in practice promotes the role of ESCOs[3] and on-bill financing but originally has only focused on EE. The scope of energy saving obligations could be expanded to also include RET in the built environment.

Further considerations

Realising EE measures versus realising RET

Business models analysed in this book implement both RET and EE measures. In the built environment, there is often not such a clear distinction between RET and EE measures as both reduce the consumption of energy from the grid or delivered by utilities. The business models 'energy supply contracting', 'making use of a feed-in remuneration scheme' and 'leasing of RET' are specifically suitable for the deployment of RET. The business models based on 'PACE financing', 'on-bill financing', 'developing properties certified with "green" building label' and 'integrated energy contracting' work for both RET and EE measures. Moreover, energy saving obligations could be extended to cover RET. 'Energy performance contracting' is especially suited for EE measures.

Generally speaking, energetically, but also from a cost-effectiveness perspective, an integrated approach is preferred which first realises energy savings and then implements RET. Such a perspective is taken by the business model 'integrated energy contracting'.

Business models versus obligations/standards

The analysis has shown that business models can play an important role in supporting an increased deployment of RET, but that they cannot overcome all barriers and are certainly not the only approach for *supporting* RET. Alternatively RET can, for example, be mandated through obligations or potentially in building codes. Examples for obligations for RET are the Spanish solar thermal ordinance and the Renewable Portfolio Standards in the US and Japan, although the latter do not focus specifically on the built environment. There is a strong rational behind implementing obligations for RET and strengthening building codes, as experience has shown that voluntary measures alone are not sufficient to drive a significantly increased deployment of RET in the built environment. The revision of the EU Energy Services Directive (ESD) for example reflects that the existing regulatory framework and incentive programmes in place are not sufficient to reach the EU's 2020 energy savings goals. However, whereas many of the business models analysed here focus on existing buildings, the potential reach of obligations for the energy performance of buildings tends to be limited to situations where a building permit is required, thus for new buildings and substantial renovations. In these situations, obligations are the 'stronger' instruments as they fully overcome the barrier of 'low priority of energy issues'. However they tend to be less flexible than market-based instruments which form the basis for many business models, and require public acceptance.

There may also be a complementary role for obligations and business models based on new financing schemes and energy contracting business models: an obligation in itself does not solve the barrier of 'high up-front costs' and 'access to capital' for the building owner or property developer who has to undertake the investment, nor does the obligation address the issue of how it is to be implemented. Thus business models could play a role in supporting building owners with providing access to capital and additional services. There may also be situations where policy makers have to decide between passing an obligation and supporting market-based solutions/business models. The current study does not allow for a general statement of what would be the best approach in a specific situation. This would require further research.

Business models and the EU 'Energy Performance of Buildings Directive' (EPBD)

The analysis undertaken in this book focuses on the contribution that business models can make to an increased deployment of RET under today's policy environment and on lessons learned from existing experience. In the future, it is expected that there will be an increased interest to support RET and EE measures in order to meet GHG reduction targets. A concrete example for tightening regulation on this subject is the 2010 recast of the EU 'Energy Performance of Buildings Directive' (EPBD), which will require among others that (1) at all major renovations, building owners adopt cost-effective EE measures, that (2) the public is provided with more information on costs and benefits of measures that improve the energy performance of buildings and that (3) by 2018 new buildings occupied and owned by public authorities are nearly zero-energy buildings. By 2020 all new buildings will be required to be nearly zero-energy buildings (EC, 2010).

There seems to be a two-way relationship between the EU EPBD and the role of business models. For some business models the implementation of the directive is a prerequisite. The business model based on rent increases after the implementation of EE measures frequently requires a certification of a building's energy performance such as those already mandated in the current version of the EPBD (EC, 2002). If EU Member States have not fully implemented this certification system, the business model based on rent increases faces challenges in its implementation. On the other hand, business models can also support the implementation of this kind of directive. The stricter requirements for new buildings and major renovations will require significant additional capital, which building owners or property developers may not have. Business models addressing the barrier of high up-front cost may support building owners in this.

There is no concrete definition, yet, of what exactly constitutes a nearly zero-energy building, but it is expected that the definition will also imply requirements of energy management and RET in buildings. Moreover, the 2010 recast of the EPBD also contains requirements for existing buildings, e.g., it requires Member States to establish requirements for the installation, adjustment and control of the technical building systems in existing buildings. Both the management of new zero-energy buildings and such requirements for existing buildings will require specialist knowledge, which building owners or managers may also not have in-house. Thus there may also be additional demand for ESCOs or similar companies to supply this expert knowledge.

Today, business models can also play a role in preparing governments and market actors for upcoming stricter regulation. The highest levels of

Figure 5.1 Crossway, which was built in 2008–2009, is a 'zero-energy' building in the UK (Photo: Oast House Architecture)

voluntary 'green' building certification may, for example, already imply that buildings are 'nearly zero-energy'. The experience gained with these buildings today provides lessons learned on costs and benefits of such high energy standards. Moreover, architects, property developers and other actors in the building supply chain, who are engaged in designing, developing and building such certified buildings today, can develop the required skills, giving them a competitive advantage for the time when these standards become mandatory.

Scope of analysed business models

The list of business models analysed in this study is not comprehensive. There are, for example, a number of additional financing schemes that address the barriers 'difficult access to capital' and 'high up-front costs'. Green mortgages, for example, allow building owners to finance energy efficiency improvements to new or existing buildings as part of or through an addition to their mortgage. The interest rates of green mortgages may be lower than market rates or a green mortgage may allow the home owner to take a larger loan than normally allowed without having to increase the down payment. This approach offers building owners access to capital at attractive interest rates and with long payback times. The additional mortgage payments are financed through lower energy bills. Thus for the home owner the basic principles are similar to PACE and on-bill financing. In addition revolving funds set up by governments with a specific focus on RET or EE measures can provide access to capital to building owners, either directly or through e.g. a municipality.

If carbon credits can be issued and monetised, these can generate an additional revenue stream which makes investments into RET/EE measures viable and can form one component of a business model. In countries which are eligible for undertaking Clean Development Mechanism (CDM) or Joint Implementation (JI) projects, programmatic CDM/JI can cover RET/EE measures in the built environment.[4] However, to date CDM/JI has not been very successful in the built environment.

It is highly likely that a number of further new and innovative business models are currently being piloted in the IEA-RETD member countries. Moreover, innovation is not limited to the development of new business models, but can also play an important role when combining existing models. The case studies of the support programme for solar water heaters in Tunisia (PROSOL) (see Appendix A.4) and of Volkswagen/Lichtblick in Germany (see Appendix A.7) illustrate that new and innovative business models may arise from a combination of elements of existing models in a way that addresses the specific barriers preventing an increased deployment of RET in the respective situation.

*Gaps in scope of existing business models: ESCOs for small
residential buildings*

The current analysis shows that the business models described here cannot
address all barriers. A specific gap identified in the analysis of the energy
contracting business models is the fact that conventional ESCOs do not work
for small buildings due to the high associated transaction costs (see Chapter
4). However, there may be significant potential for companies, e.g. individual
installers or groups of them, in offering less extensive or more standardised
energy services to individual households than conventional ESCOs deliver for
larger buildings. Individual home owners tend to have relatively little interest
and knowledge in energy related issues, but comprehensive service packages
could take away the 'hassle factor' and allow building owners to outsource
energy services. In the Netherlands, such approaches for improving the energy
performance of existing buildings are for example stimulated by the government's 'More with Less' (Meer met Minder) programme (see example
Wonen++ in Box 5.1). However, there is little comprehensive information on
the extent to which such energy services are already offered across IEA-RETD
member countries. There is a need for further research in this area, and a role
for policy makers to support innovative, new approaches (see Chapter 6).

Box 5.1 Case in point: Wonen++ concept – an example of small-scale energy services

The 'Wonen++' concept of the Dutch energy company Eneco is an example of
small-scale energy services. With this business activity Eneco supports the
realisation of energy saving measures and renewable energy in households.
Eneco offers several organisational and administrative services, like an energy
audit, planning and installation. It does not invest for its clients nor does it engage
in performance-based contracts; clients have to finance the investment costs
(net of subsidies) themselves. However, Eneco offers services related to
financing, like searching for and requesting government financial support (soft
mortgages, personal loans, or subsidies). Wonen++ is a member of the Dutch
government's 'More with Less' programme, which allows Wonen++ to offer a
small subsidy to their clients.

According to Vethman (2009) this was one of the very few small energy
service companies found to offer services to private home owners at the time
of that study. An important reason for a lack of supply of, and demand for,
energy services in the residential sector are not only financial barriers, but
equally important non-financial barriers like a lack of awareness, hassle and the
complexity of the energy services concept.

Box 5.2 Case in point: cost-effectiveness of RET may vary

Example of a solar water heater

The cost-effectiveness of RET in the built environment depends on many factors including the conditions for financing up-front capital investments, technology specific energy yields, and the cost difference versus conventional energy generation.

Consider a typical 2.4 m² collector, thermo-siphon solar heating system with storage volume of 150 litres, assuming average investment costs of around €650/m² installed, VAT (20%) excluded (IEA, 2007). Thus the total investment costs including VAT are €1872. When building owners pay the investment for the solar water heater from their own funds, the opportunity costs will equal the lost interest in a savings account, e.g. 2 per cent (not taking the risk of the project into account). In this case, the annualised investment costs are €146 per year. The energy yields for a domestic hot water solar heating system with high solar fraction in Southern Europe, such as France, are 800 kWh/m² per year (IEA, 2007). The solar heating boiler produces 1920 kWh per year. If this heat were produced with a conventional electric boiler with a conversion efficiency of 90 per cent, heating the same amount of water would require 2133 kWh electricity. In France, end-user electricity prices in July 2011 were 0.14 €/kWh incl. all taxes. Thus, in this case in a Southern European country with high solar fraction, the solar heating boiler would save about €300 of electricity costs a year, implying a net saving of about €150.

If the same solar boiler was installed in Northern Europe, such as in the Netherlands or Germany, the energy yield would be 250 kWh/m² per year (IEA, 2007). In this case, the solar heating boiler produces only 600 kWh per year. If this heat was produced with a conventional gas fired boiler with a conversion efficiency of 75 per cent, as is common in e.g. the Netherlands, producing the same heat would require 800 kWh of gas. The end-user gas price in the Netherlands or Germany in July 2011 was about 0.07 €/kWh incl. all taxes. The solar water heater saves €56 of costs for gas a year. In this case, the annualised capital costs of the solar water heater are higher than the cost savings.

If alternatively a building owner borrows the money for the investment commercially, interest rates and administration costs for concluding the loan agreement are higher. If the solar boiler was for example financed through the mortgage on the building, the interest rate could be around 5 per cent and there are no additional costs for the loan agreement. In this case, the annualised capital costs would increase to €180 per year. If the solar boiler is financed through a personal loan, the interest rate could be up to 10 per cent and the administrative costs for concluding the loan agreement would probably be relatively high compared to the small amount borrowed. However, an ESCO could aggregate demand for the equipment and the financing of a large amount of solar water heaters, which would lower the administrative overhead considerably.

Example of solar PV

Solar PV has traditionally been a relatively expensive renewable energy technology. However, today, in some Southern European countries, grid parity (i.e. production costs equal end-user electricity prices) is almost reached for solar PV, e.g. in Spain (Jumanjisolar, 2011). An average household in Spain uses 3000 kWh electricity every year. Electricity costs are €521, assuming an electricity price of €0.1737 incl. energy tax and VAT per kWh. To produce this electricity by solar PV, 2.54 kilowatt-peak (kWp) of solar panels would be required, as solar PV delivers 1400 kWh per kWp in Spain. The total investment costs for such a system would be €7586 (at €3540 per kWp). With a lifetime of 25 years this amounts to an annual payment of €538, assuming a 5 per cent interest rate on the capital investment. According to this example, in Spain the electricity generation costs of solar PV are almost the same as the electricity prices paid by consumers.

For countries in Northern Europe more solar PV capacity is needed to produce the same 3000 kWh per year. In the Netherlands, solar PV delivers for example 800 kWh per kWp. The total investment costs for a sufficiently large system would be €17,280 (assuming €4608 per kWp including installation). With a lifetime of 25 years this amounts to annualised costs of €1226 per year at 5 per cent interest rates. The same amount of electricity from the grid would cost €540 (electricity price of €0.19 per kWh incl. all taxes).

These calculations assume constant prices for conventional energy in the future. RET will become more cost effective if the prices for fossil fuel rise and/or the prices for RET decrease. While for the RET discussed there, cost-effectiveness still depends largely on the background situation, some EE measures, such as improving insulation of walls, roofs and windows to a certain level, are almost always cost-effective.

[All energy prices based on www.energy.eu, accessed in July 2011]

Potential for further research

Based on the analysis undertaken in this book, there are several open questions and topics for further research on business models for RET and EE measures in the built environment and on the role of policy makers in supporting these models.

On energy services

As stated above there is a need for further research on innovative approaches for expanding energy services to smaller residential buildings. One research direction could be to look into the added value for building owners, e.g. into the costs and benefits of service providers realising measures compared to building owners undertaking them themselves. This would enable more insights into which situations energy services for individual home owners are suitable. Another approach would be to further analyse how such energy services can lead to business cases for service providers, e.g. by standardising the offering.

There is also a need for further research into the role of energy suppliers in offering energy services to their customers, e.g. which approaches work best and why, can they be standardised, and can the (financial) attractiveness of these approaches for the consumers be improved? Based on this, recommendations for policy makers, e.g. for the design of energy savings obligations, could be derived.

On 'green' building certification schemes

For the US market there are already a number of studies on certified 'green' buildings analysing savings in operating costs and additional benefits such as increased well-being and productivity of building users, as well as market value compared to non-certified buildings. However, outside the US, for the European and Japanese market, there are no such major studies, yet (Nelson et al., 2010). Certification bodies, and property developers and owners with a large portfolio of certified green buildings, could take a proactive role in collecting and publishing data on the performance of certified buildings. However, to ensure credibility the studies themselves would best be undertaken by external organisations.

Innovative financing schemes

There are several innovative financing schemes that have not been covered in this study, such as 'soft leases' (leases with a lower interest rate, similar to soft loans), financial guarantees (e.g. a guarantee scheme for green mortgages) or revolving funds (e.g. for EPC projects). Moreover, there is scope for increasing the understanding of how financial institutions can be better involved in offering financial products for financing energy improvements in buildings.

Role of business models versus obligations

As discussed above, there is a strong rational behind implementing obligations for RET and strengthening building codes, as experience has shown that voluntary measures alone are not sufficient to drive a significantly increased deployment of RET in the built environment. In order to optimally tailor policy interventions, there is a need for a better understanding in which situations business models can play a role as a real alternative to standards/obligations.

Notes

1 These business models are: the ESCO models, on-bill financing, PACE financing, leasing and the business model based on the ability to charge higher rents to tenants.
2 In the EU ESCO market for example, the lack of sufficient projects was assumed to be the main bottleneck for market growth in the past, rather than the availability of financing, i.e. not enough building owners had an interest to involve an ESCO.
3 In Denmark for example, energy services are implemented because energy saving obligations for energy suppliers may only be implemented by third party energy service companies.
4 See for example, the proposed JI programme 'Energy Efficiency Programme in Buildings' by BOS Bank in Poland (http://www.netinform.de/KE/files/pdf/Bos_EnEff_PoA-DD_2010_04_16_final_4.pdf).

Chapter 6

Recommendations for policy makers and market actors

This chapter provides recommendations for policy makers, building owners and market actors active in the built environment based on the analysis in this study.

Recommendations for policy makers

Active and strong role of policy makers required

Even though a number of business models such as the energy contracting models do not require direct policy intervention, some others do. In any case, policy makers can support even the business models that do not require direct intervention. New and innovative business models can help to exploit the potential of sustainable energy in the built environment. However, the analysis indicates that voluntary measures in support of the models are not sufficient to explore the full potential of RET as other important barriers remain, such as 'low (or no) returns on investment' or 'low priority of energy issues'. The business models described in the book are therefore not expected to lead to a significantly increased deployment of RET without further supporting measures. Moreover, in the past, regulation has been the strongest driver for increasing energy efficiency of the building stock. This points towards an active and strong role of policy makers.

When supporting business models, the first step is to consider which RET are cost-competitive compared with 'traditional' energy sources in a specific jurisdiction, as most business models will generally only lead to a deployment of cost-competitive technologies. The second step is to determine the most suitable business models given the market segment in which RET is to be deployed. The analyses of the business models evaluated in this study provide guidance for this: Table 5.1 can be used to determine which business models would work in a specific market segment. Chapter 4 supports this choice of model by providing more details on each business model.

The following first provides recommendations for business models which lead to a deployment of cost-effective technologies, and later addresses approaches for situations where RET are not (yet) cost-competitive.

Supporting business models in existing large residential, commercial and public buildings

Existing buildings are a challenging market segment for increasing the deployment of RET and EE measures as the market segment is difficult to reach with building codes and obligations, pointing to an important role for business models. To support business models in this segment, policy makers can:

- support energy contracting business models (see Box 6.1 for details on potential support measures).
- facilitate leasing of energy equipment by resolving regulatory barriers and potentially supporting banks in offering 'soft leases' similar to existing 'soft loans'.
- explore the potential for a programme similar to PACE financing such as the programme undertaken in Melbourne, Australia (see Appendix A.3).

Box 6.1   Supporting energy contracting business models

Energy contracting business models are the most complex of the models analysed in this book, as ESCOs offer comprehensive service packages to their clients. In contrast to many of the other business models, ESCOs do not directly depend on policy or financial support but are market-driven. However, there are still various approaches for governments in supporting energy contracting business models, e.g. by:

- financially supporting the establishment of independent third party organisations, e.g. energy agencies, which act as market facilitators and as project facilitators between potential customers and suppliers. Such organisations have proven to be key for supporting strong market growth, but are still lacking in most countries.
- creating model contracts and monitoring and verification standards, and setting up national or regional 'public knowledge centers' in order to ease access to information. Transparency and trust in the market could be increased by providing, for instance, lists or registers of energy services offered, and performing random quality checks.
- implementing a range of instruments in order to facilitate access to financing (e.g. guarantee funds, low interests loans using revolving funds), which is frequently a challenge for ESCOs.

- changing procurement rules for public buildings to allowing and requiring decision makers in public buildings, including public housing corporations to procure equipment according to lowest life-cycle costs and allowing them to enter into long-term contracts. In addition, rules can be adapted so that RE and EE investments under an ESCO contract are not treated as public debt.
- developing and financially supporting models specifically customised for households and small and medium enterprises would significantly widen the market potential of ESCOs, which is currently mostly limited to large public, commercial and residential buildings or to pools of buildings. Governments may play a role in encouraging and promoting innovation and creativity especially among small and medium sized ESCOs by supporting research on innovative models and market actors, by supporting project development including model documents or by supporting pilot project implementation. A standardisation of products and contracts offered by ESCOs could, for example, significantly reduce transaction costs.

In most countries some of these measures need to be taken by national government, e.g. changes in procurement rules. For other measures, there is also room for action by local or regional governments, e.g. through the creation of third party market facilitators (see for example the German and Austrian energy agencies (Energie Agenturen)).

Supporting business models in existing small residential buildings

PACE financing and on-bill financing can target this market segment, as can programmes undertaken by utilities in the frame of fulfilling energy savings obligations. All of these help to overcome the barriers of 'high up-front costs' and 'lacking access to capital'.

- Before supporting any of these business models, policy makers should consider which of the models would be suitable for the specific situation (see Box 6.2).
- Utilities can have a key role, as they have direct access to building owners, i.e. the decision makers for deploying RET/EE measures in existing buildings. Policy makers can mandate, e.g. through energy saving obligations, or strongly incentivising utilities to take a strong role, for example by restructuring a utilities revenue structure in a way that rewards energy savings and installation of RET in buildings.

- To facilitate on-bill financing programmes, policy makers can take additional measures such as:

  - clarifying legal issues around liabilities created through on-bill financing programmes
  - partnering with utilities in providing access to capital for the programme, e.g. through revolving funds
  - offering the opportunity to combine the programme with subsidies to enable the installation of a wider range of RET/EE measures.

- PACE financing requires a change in legislation. As it is a new and innovative business model with which there is only limited experience outside of the US, it is advisable to start implementation on the scale of pilot programmes to gain experience on whether the approach works in the country and targeted market segment and to derive lessons learned for a potential scale-up of the programme (see also example in A.3). In the US and Australia, the required legislative changes need to be undertaken by state governments, but the level of decision making may vary among countries.
- National or regional governments can facilitate access to capital for the entities who undertake a programme, e.g. by establishing revolving funds which utilities or local governments can access.
- To further support the uptake of RET, RET could be included in energy saving obligations for energy suppliers.

Box 6.2 Considerations for supporting business models based on new financing schemes

From the point of view of the building owner, in most of the business models based on new financing schemes the investment costs for the RET/EE measures are spread over the lifetime of RET, so the building owner does not need to provide the up-front investment. However, these business models differ with regards to the involved actors. It is not possible to make a general statement which model is the best, as this depends strongly on the local circumstances. For policy makers aiming to support one of these approaches relevant considerations are:

- Which actor has the financial means to offer support schemes? In economically difficult times local governments may, for example, face challenges in facilitating access to capital. Banks are generally best suited to provide access to capital.

- Who has an intrinsic interest in facilitating a financing scheme for RET/EE measures? Support for RET/EE measures may, for example, not be in the direct interest of utilities or banks.
- Who has the technical capacity to implement such a programme?
- Who has access to the decision makers for investments in RET/EE measures, i.e. the building owners?
- Out of the business models based on new financing schemes, PACE financing and on-bill financing have the advantage that the debt for the RET/EE measures may stay with the property or electricity meter. This is especially important in markets where buildings change owners frequently. However, both schemes can be complex in terms of administrative and regulatory efforts required.

Supporting business models in rental buildings

It is recommended that governments introduce legislative changes that allow property owners to adjust the level of rent after undertaking investments in EE, if there is significant energy savings potential in the regulated rental market. A few EU Member States (the Netherlands, France, Germany, the UK, Italy and Sweden) have already gained experience with this.

Supporting business models in new buildings

In contrast to existing buildings, in new buildings, EE measures and RET can be mandated through building codes and obligations. In the EU, the recast of the EU Directive on Energy Performance of Buildings (EPBD) is expected to be the main driver for moving new buildings to becoming more energy efficient. 'Green' certification may complement regulation and building codes by supporting innovation by setting standards for highly energy efficient and environmental buildings and increasing transparency on their performance and benefits. Such 'green' certification schemes can be supported by:

- applying 'green' certification to public buildings or even making them mandatory for certain types of (public) buildings as is currently happening in the UK.
- working together with certification bodies to ensure that the requirements for voluntary and mandatory systems are harmonised.

In addition, the recommendations for energy contracting business models (see Box 6.1) apply to new buildings as well.

Supporting non-cost-effective technologies

Business models themselves cannot address the barrier of 'low (or no) return on investment'. Today, there are already cost-effective opportunities for deployment of RET and EE measures (see e.g. Box 5.2). For technologies where this is not (yet) the case, business cases may be based on supporting policy measures such as feed-in remuneration schemes (see Chapter 4), direct subsidies, fiscal measures like tax breaks, or interest rate subsidies.

The advantage of feed-in remuneration schemes over direct investment subsidies and soft loans is that feed-in schemes directly incentivise the production of renewable energy and not just the installation of the technology. Successful feed-in schemes are stable and predictable, so policy makers should change categories and tariffs in a feed-in scheme only when technology cost reductions or changes in energy prices demand doing so while giving sufficient lead time and full transparency on reasons.

Overcoming other barriers

The business models described in this book do not address all barriers for an increased deployment of RET in the built environment. Some barriers, e.g. the barrier of 'cumbersome building permitting process' or 'inappropriate mortgage assessment', cannot be addressed by business models at all, but require changes in regulations instead. Removing such barriers can be a precondition for the business models to be successful. Thus in addition to supporting or enabling business models, additional policy measures are required to overcome these barriers.

Using policy packages to address various barriers at once

Successful business models often consist of combinations of the models described in this book, simultaneously addressing several barriers. To support such approaches, a combination of policy measures or policy packages is needed. Such policy packages could, for example, combine energy saving obligations for energy companies with incentives such as subsidies, soft loans, or loans with guarantees, and support for the provision of tailored information through e.g. energy audits. Or there could be a number of policy measures supporting energy performance contracting, such as applying EPC in public buildings and supporting market development through a government administered energy agency. A differentiation of the rental price system for social housing could be combined with a good energy label system. Depending on the specific background situation in a certain jurisdiction and market segment, other combinations are possible.

Recommendations for building owners

For all building owners

The business models described here offer attractive opportunities to building owners by facilitating access to capital, overcoming the barrier of 'high up-front costs' and offering a wide range of energy related services and thus also addressing some non-financial barriers for building owners, such as the 'hassle factor'. Building owners are encouraged to explore which business models may be advantageous in their situation.

For owners of existing and new large commercial and residential buildings

ESCOs present a special opportunity as they allow for outsourcing technical and economic risks of RET and purchasing a comprehensive service package. Moreover, energy contracting business models may spread the investment costs of RET/EE measures over the lifetime of the equipment, allowing for investment in RET/EE with minimal capital outlay.

For public building owners

Public building owners play a special role, as they can serve as a means to drive the implementation of government targets for the deployment of RET and EE measures in the built environment. The decision to deploy RET/EE measures can only be taken by building owners. For the public housing stock, however, these decisions are under the direct influence of governments. In the EU, publicly owned or occupied buildings represent about 12 per cent of the entire building area (Ecorys et al., 2010), indicating that the impact of government action regarding public buildings is significant. Public buildings can serve as a 'role model' and governments should be proactive in applying suitable business models. Public building owners can for example:

- apply certification with voluntary 'green' building labels to new buildings and during substantial renovation of existing facilities
- directly support ESCO business models by using these models in the public building stock. This may require a change in public procurement rules.

In addition to the direct impact on its housing stock, public building owners can make other actors aware of the potential through demonstration and dissemination of the impacts of these business models. Supporting new and innovative business models in public buildings poses a unique opportunity for local governments to become active in increasing the deployment of RET in the built environment, as the responsibility for the public building stock usually lies at the local level.

Recommendations for other market actors

Recommendations for market actors actively involved in business models

Market actors are recommended to analyse markets well to ensure that there are no important stakeholders that can block a business model. In the example of the 'PACE financing' model in the US, the interests of all stakeholders were for example not taken properly into account, so that the residential scheme was stalled by an intervention of the mortgage reinsurers.

Recommendations for ESCOs

ESCOs can take specific actions to support the development of the energy services market, which in most countries is still relatively young and has little structure, for example by:

- supporting market facilitators and independent consultants who prepare projects for public and private building owners and put out calls for proposals
- helping to develop and agreeing on model documents and procedures
- establishing a sector organisation which provides market data and participates in policy development
- being transparent regarding business models used to gain trust of the potential customers
- participating in conferences or other public meetings to inform market actors and create awareness on energy related issues
- further developing business models (e.g. for individual home owners).

In addition, the analysis also showed that often business models are most successful when they are based on partnerships between actors with complementary expertise and resources, e.g. regarding access to capital, technical expertise and access to the clients/building owners. An example of such a successful programme is the Tunisian PROPOSOL scheme (see Appendix A.4), which brings together key actors in the sector including the state electric utility, commercial banks, the suppliers of solar water heaters, and building owners or occupiers. Actors directly involved in business models are recommended to explore partnerships with others.

Appendix A Case studies

A.1 LIG, Austria – integrated energy contracting pilots

(Business model: integrated energy contracting; see page 41)

The Landesimmobiliengesellschaft Steiermark (LIG) (State of Styria owned real estate company) administers and manages more than 420 buildings in Styria. About 200 buildings with an overall area of more than 600,000 m² are owned by LIG. To our knowledge, LIG is the first institutional building owner that has systematically applied the concept of integrated energy contracting. Recently, LIG's IEC activities have been recognised with the Energy Globe Styria Award 2009.

In 2007/08, LIG made the first Europe-wide IEC call for tenders for five buildings with a net floor area of approximately 11,000 m². In 2009, contracts for pool 2, which consisted of three properties with altogether 20,000 m², were procured and implemented. Contracts for another pool of buildings are under preparation.

The original motivation of LIG was to substitute heating oil as far as possible with energy carriers that are renewable. However, in the course of the project development, the objectives of LIG's call for tenders for IEC contracts were extended as follows:

1   Implementing demand side saving measures with payback times of less than 15 years in the fields of building technology, building shell and user behaviour, and improving the energy indicators of the buildings;
2   Comprehensive refurbishment of all oil-fired heating equipment;
3   Reducing CO_2 emissions (which implies a change of energy carriers) and minimising overall energy cost.

Good practice example Retzhof

The Retzhof of LIG is a complex of buildings consisting of a castle from the sixteenth century as well as two seminar and guest houses from 1960 and 2009 with an overall useful area of about 4,000 m². The buildings are used as a hotel and seminar house.

Figure A.1 'Schloss Retzhof': seminar house of the Province of Styria

The initial situation before refurbishment and the construction of the new building can be summarised as follows: high energy costs, an inefficient natural gas boiler and no insulation of the castle building (protection of a historic monument). The old boiler house had been demolished to make room for the new guest house, including the new heating centre. Energy consumption amounted to approx. 185 kWh/m²/year.

The building owner had the following goals for the refurbishment:

1   Replacing the old boiler installation;
2   Outsourcing energy supply and financing the investments;
3   Reduction of energy demand, CO_2 emissions and costs through demand side saving measures.

The project was implemented with the support of Grazer Energieagentur GmbH as an integrated energy contracting model. Figure A.1 shows a schematic representation of the contractual relationships and cash flows of the project. Central issues were the combination of energy efficiency measures and supply of useful energy, and using specific quality assurance instruments to substitute the EPC savings guarantee. The ESCO contract was awarded in a combined competition of prices and solutions in the course of a two-phase negotiation procedure.

From the building owner's perspective, some important experiences and innovative approaches of the project can be summarised as follows:

1   The combination of energy efficiency and supply of useful energy within the IEC model basically works.
2   From the building owner's perspective, a coordinating and controlling function is necessary even if the ESCO acts as a general contractor, especially if other building construction projects are simultaneously carried out as in-house implementation (in this case the construction of the new guest house).
3   The development of comprehensive energy (efficiency) projects requires committed facilitators and a long time horizon.

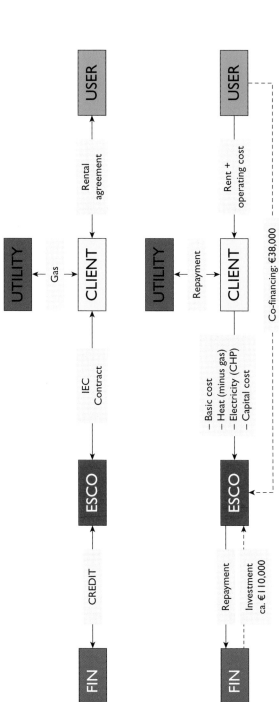

Figure A.2 Schematic representation of the contractual relationships (upper graph) and cash flows (lower graph) of the integrated energy contracting (IEC) pilot project 'Schloss Retzhof'

4   Thanks to co-financing of the investments by using funds provided by the user of the building, the ongoing capital costs could be reduced by approximately 30 per cent.

5   The ESCO invested in the CHP plant upon its own risk. Re-financing will be done by selling electricity to the building owner during the contractual period.

These results apply subject to a systematic monitoring and verification in the course of the annual auditing of the buildings.

A.2 Berkeley FIRST – the first PACE financing programme

(Business model: PACE financing; see pages 73, 79, 110)

In 2008, the City of Berkeley launched the first ever PACE financing programme, called Berkeley Financing Initiative for Renewable and Solar Technology (FIRST). The programme aimed at testing the viability of PACE financing. The small pilot programme focused exclusively on solar PV installations to keep the process simple. However, home owners were required to also undertake basic energy efficiency improvements. Funding was based on micro-bonds issues by the City of Berkeley and purchased by Berkeley's financial partner. The participating home owners are repaying the financing of their solar PV systems via their property tax bills over a period of 20 years. Table A.1 shows a sample calculation of investments and costs for a home owner.

Out of the 40 slots which were available under the pilot programme, 13 projects were completed for a total of USD 336,550 of financing via micro-bonds. The interest rates offered to property owners participating in the pilot programme were relatively high at 7.75 per cent, which was equal to 3.25 per cent above the 10-year U.S. Treasury Note at that time. The level of interest rates was considered a barrier by many of the property owners applying to the programme who withdraw their application: some of them still installed a PV system but used cheaper financing options.

The Berkeley FIRST programme provided valuable insights into how PACE financing can be applied in practice. For the 13 projects realised under the programme, an estimate of the project economics showed a negative net present value for the property owner in 9 of the 13 projects. The savings for the home owner are generated through 'net-metering', i.e. lower electricity bills due to auto-consumption of the generated electricity. Solar PV is generally still a relatively expensive renewable energy technology, thus the negative Net Present Value (NPV) of the initial projects is not surprising. Since 2008, the prices for solar PV systems have dropped significantly, implying that the same programme would be more cost-effective today. Moreover, it is expected that larger programme sizes could reduce the financing costs, and

Table A.1 Sample calculation for the installation of a solar PV plant on a residential building under the pilot Berkeley FIRST PACE financing programme

Cost of the PV installation	USD 28,077		
California Solar Initiative (CSI) subsidy	USD –6,108		
Administrative cost	USD 600		
Total PACE financing		USD 22,569	
Interest rate		7.77%	
Annual tax payment by the property owner to repay the loan			USD 2,199

Note: For the building owner the additional tax payment is (partly) offset by lower electricity costs.

Source: Brown et al., 2010.

that future programmes could also include energy efficiency measures, which would be expected to raise NPVs of the projects (based on City of Berkeley (2010), Brown & Conover (2009), RAEL (2009)).

A.3 Financing commercial building retrofits with the help of an environmental upgrade charge in Melbourne

(Business model: PACE financing; see pages 73–4, 110, 124, 126)

In 2010, the City of Melbourne, Australia, initiated a new programme for financing energy retrofits in commercial buildings. The financing approach is similar to PACE financing and has some characteristics that could be replicated and scaled up in other parts of the world.

The programme is part of Melbourne's Zero Net Emissions by 2020 Strategy, and aims to retrofit 1,200 existing office buildings to decrease the use of energy, water, and the generation of GHG emissions. The financing structure is based on a newly introduced environmental upgrade charge (EUC), which was included in an amendment to the City of Melbourne Act passed by the Victorian Parliament in September 2010. This amendment enables the city council to enter into so-called environmental upgrade agreements (EUAs) with commercial property owners who look for up-front financing for environmental retrofit projects, and with financial institutions which are willing to fund these retrofits. In contrast to PACE financing programmes, in this case the building owners themselves are responsible for arranging the financing terms with financial institutions. When the EUA is approved, the bank will lend money to the building owner for undertaking the retrofit project. The building owner then repays the investment and interest via an ongoing environmental upgrade charge (EUC) levied by the city council who passes these payments on to the lender.

The programme does not target single buildings, but portfolios of properties. In its initial phase, top-tier non-residential property owners are eligible for the programme that have an investment grade credit rating, own 10 or more properties with a total of more than 5,000 m² of floor area and have annual energy costs of at least $500,000. These criteria will help to lower administrative costs and improve the likelihood of long-term success.

The advantages of the owner-arranged financing structure are that building owners are able to secure more attractive interest rates than the municipality would be able to and that lending terms can be structured specifically to the project. However, the approach would likely have to be adapted if the programme was expanded to smaller property owners. The EUC and involvement of the city council provide additional security to the participating financial institutions. Applicants must show that the planned retrofits are expected to achieve energy savings of at least 20 per cent. This requirement was defined in order to promote comprehensive retrofits instead of simple add-on solutions. It is expected that the initial experiences will provide lessons learned for a future expansion of the programme.

(Based on Institute for Building Efficiency, 2010c)

A.4 PROSOL: supporting market growth of solar water heating in Tunisia

(Business model: on-bill financing; see pages 80–1, 86, 117, 130)

In 2005, the Tunisian Ministry for Industry, Energy and SMEs and the National Agency for Energy Conservation (ANME) together with UNEP launched the programme PROSOL, which included an innovative combination of support mechanisms to create successful business models for financing, supplying and installing solar water heaters (SWH) (see Figure A.4 for the organisational structure of the PROSOL business model). Financial incentives were targeted at reducing the high up-front cost of capital for the installation of equipment and at improving access to capital by reducing interest rates and organising the repayment of loans for the SWHs through an on-bill financing mechanism. The financial mechanisms consisted of three elements:

- A capital cost subsidy provided by the Tunisian government (and partly by UNEP through the Mediterranean Renewable Energy Center (MEDREP)) to customers for 20 per cent or more of the initial cost of the SWH.
- Reduction of interest rates through an agreement with commercial banks to charge lower interest rates due to the reduced default risks within PROSOL and through an additional interest rate subsidy. The latter has been progressively phased out over a period of 18 months.
- An on-bill financing mechanism where customers who install a SWH repay the loan via their electricity bill over a period of 5 years.

The financial instruments were accompanied by a series of additional measures including supply-side promotion, development of a quality control system for SWH equipment, an awareness raising campaign and a capacity building programme with financiers to raise their understanding of RETs. The programme brings together key actors in the sector such as the state electric utility STEG (Société Tunisienne d'Electricité et du Gaz), commercial banks which provided the most favourable loan conditions determined in a bidding process, the suppliers of SWH which include local manufacturer and importers, and customers who install the equipment.

An interesting component of the PROSOL scheme is that initially it relied heavily on the initiative of suppliers of SWH systems who took the role of indirect lenders of money for their customers, the home owners. Suppliers are accredited for participation in the programme by the National Agency for Energy Conservation (ANME). Customers have to be clients of the state electric utility STEG to participate in the programme. When customers decide to purchase a SWH under the programme, they need to sign an agreement form where they commit to paying back the loan via their electricity bill. Customers only need to pay a small amount in cash, e.g. 10 per cent of the cost of the SWH. Under the first phase of the programme, the supplier guaranteed the loan for the SWH with the local bank. The bank did not have any direct contact with the customers, as customers paid the loan over 5 years through their bi-monthly electricity bills issued by STEG.

The innovative scheme provided various additional levels of security and benefits for involved actors by aligning their interests. Through the installation of the SWH system, customers profit from lower electricity bills which compensate for the additional payment on their electricity bill. Customers also get additional security on the performance of the SWH, as the supplier guarantees for the loan with the bank and thus has an interest in maintaining the functioning of the system. The bank gets additional security because repayment of the loan is automatically recovered via the electricity bill.

Figure A.3 Solar thermal water heater installed in Casablanca, Morocco (Photo: Myriem Touhami, UNEP)

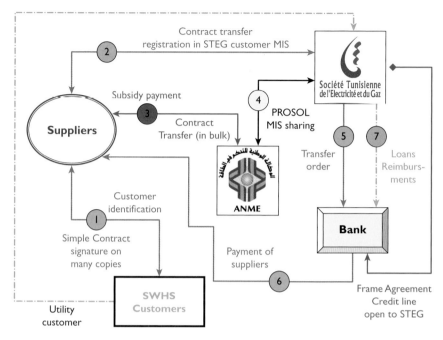

Figure A.4 Organisational structure of the PROSOL business model (source: Missaoui & Mourtada, 2010)

Moreover, in the case of default, the bank can take actions against both the supplier and (via STEG) against the consumer. In the worst case, STEG is even authorised to suspend electricity supply to the customer. For suppliers and installers of SWH the scheme has opened significant new business opportunities as it contributed to a rapid growth in SWH installations.

While on average only about 10,000 m^2 of SHW systems were installed annually in the period between 1997 to 2004, installations increased to about 90,000 m^2 in 2010 (see Figure A.5). Overall, by 2010, 95,000 SWH systems with a total capacity of 285,000m^2 had been installed under PROSOL. The number of companies selling SWH systems increased from 8 in 2004 to more than 40 in 2009, while the number of qualified installers grew from about 100 to 1000 in the same time.

A disadvantage of the initial system was, however, that successful suppliers took on large amounts of debt by guaranteeing the loans for their customers. In the second phase of the programme this was changed and customers now deal directly with the local banks.

(based on Usher, 2010; Menichetti & Touhami,
2007; Missaoui & Mourtada, 2010)

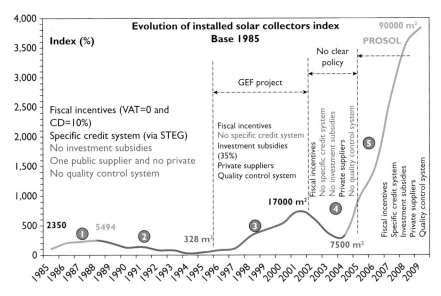

Figure A.5 Growth in newly installed solar thermal collectors in Tunisia between 1985 and 2010 (source: Missaoui & Mourtada, 2010)

A.5 Greenchoice: solar supply contracting in the Netherlands

(Business model: making use of a feed-in remuneration scheme; see page 47)

The Dutch energy company Greenchoice supplies 100 per cent renewable energy. It recently launched a plan for a pilot project 'Zonvast', in which Greenchoice invests in solar panels and markets these to around 500 home owners. Home owners can request a solar PV panel on their roof installed by an installation company partnering with Greenchoice. The home owners will not own the solar panel, but will pay a fixed electricity price to Greenchoice for using the electricity produced. Electricity not used is fed into the electricity grid, for which the home owner receives no payments.

To join the project home owners need to meet certain technical requirements. There is a risk for home owners when they decide to move, as the contract remains binding. The contract could then be sold to the new owner, or the owner could buy the solar panels and sell them along with the house, or the owner could move the solar panels to his new house.

Greenchoice already operated this business model for business customers ('Yellow step'), proactively making use of the Dutch feed-in tariff SDE.

Data:

- A solar panel installation includes 8 to 10 solar panels per roof.
- Expected electricity produced is approximately 1,950 kWh per house-hold a year (a little more than half of the average electricity consumption of a Dutch household).
- Each installation costs Greenchoice €8,000. The expected total invest-ments costs of the whole project are €4 million.
- There is a relatively small investment subsidy available to Greenchoice.
- The contract obliges a home owner to pay a fixed electricity tariff of 0.23 €ct per kWh_e for a period of twenty years. This seems a reasonable price for electricity compared to other Dutch energy companies.[1]

(Based on Greenchoice, 2011)

A.6 Adaptation of the rental price evaluation system in the Netherlands

(Business model: building owner profiting from rent increases after the implementation of energy efficiency measures; see pages 67–8)

This case study describes the adaptation of the rental price evaluation system in the Netherlands to reward energy improvements. The scheme is intended to stimulate dwelling owners to invest in energy efficiency improvements, which they are allowed to undertake with consent of the tenant. Investment

	Points		Approx. corresponding rent (euro/month)	
	Family house	Multi-family house	Family house	Multi-family house
Label A++	44	40	198	180
Label A+	40	36	180	162
Label A	36	32	162	144
Label B	32	28	144	126
Label C	22	15	99	68
Label D	14	11	63	50
Label E	8	5	36	23
Label F	4	1	18	5
Label G	0	0	–	–

Figure A.6 Valuation of the energy-performance score (expressed in points) depending on the energy label of the building (Source: Tigchelaar et al., 2011)

costs for these improvements are borne by the building owners, in this case the housing corporations. In turn, owners aim to earn their investments back over the lifetime of the measures by setting higher rental prices.

In the Netherlands, the rental price evaluation system determines the rental price for houses and apartments rented in the social housing sector, which is a large part of the total rental market in the country. A single-family house or apartment receives points for certain aspects such as the number of toilets or the existence of a garden or balcony. After the adaptation of the system, energy performance becomes one of the evaluation criteria, i.e. a more energy efficient dwelling gets more points. Awarded points determine the maximum rental price to be charged. Assuming an average rental price per point between €4 and €5 (WS Wonen, 2010) the difference in monthly rental price between a G and an A labelled dwelling can reach up to €198. Figure A.6 shows how the additional rent that a building owner may charge differs according to the energy label of the building.

For tenants, the aim of the regulation as set by the government is to ensure lower living expenses as the scheme covers the social housing sector. Thus the decrease in energy costs due to energy efficiency measures should out-weigh the rental price increase. Furthermore, landlords are only allowed to charge a higher rent when a new tenant moves in (Eerste Kamer, 2011).[2] And rental prices are only allowed to rise when the effect of the energy efficiency measures has been proven (Woonbond, 2011) to ensure effectiveness of the regulation. A solution for the requirement to acquire consent of tenants can be the use of a 'living expenses guarantee' through which the change of living expenses level after renovation is contractually guaranteed for collectively organised tenants (Aedes, 2009). A disadvantage of this guarantee for housing corporations are the associated transaction costs for the assessment and monitoring of energy savings.

Example cash flow calculation and profit calculation

The impact of the proposed adaptation of the rental price evaluation system in the Netherlands has been evaluated and described by Tigchelaar et al. (2011). Figure A.7 from this evaluation illustrates for an average dwelling of different energy label classes the yearly benefits and costs for an average tenant and landlord.

The example calculation shows that:

- For landlords there will be a small annual net cost, although housing corporations are expected to perceive these as acceptable. The calculations indicate that reaching larger energy improvements (i.e. better energy label classes) implies lower yearly costs for landlords. This is aimed to stimulate higher investments by landlords.
- For the average tenant there will be yearly net benefits. However, for the tenant larger energy improvements lead to lower cost savings.

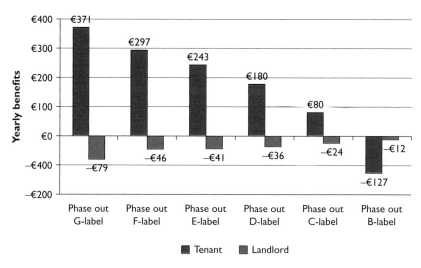

Figure A.7 Net costs and benefits of renovation for landlords and tenants, under the adapted rental price evaluation system (Source: Tigchelaar et al., 2011)

Furthermore, Aedes (2011) estimates that 100,000 Dutch dwellings with a bad energy performance will receive a lower rental price from this regulation. For these dwellings housing corporations receive lower income.

A.7 Market introduction of small and micro-CHP systems

(Business model: leasing of RET; making use of a feed-in scheme; see pages 93–4, 106, 117)

Across different countries, companies have used a model based on leasing of equipment to end-use customers or similar approaches in order to introduce small and micro-CHP systems to the market. The following presents two examples.

Tokyo Gas and Osaka Gas (Japan)

Japan is one of the leading countries with regards to the development and installation of micro-CHP systems (IEA, 2008b). The installation of micro-CHP systems in residential houses is strongly driven by large gas utilities which are competing for market share in the residential heat market against electric utilities. The electricity companies promote households working completely without the use of natural gas, which has been the traditional energy source for hot water generation (Nishizaki, 2008). The introduction of micro-CHP systems started with systems based on internal combustion engines, such as the Ecowill model, but gas utilities such as Tokyo Gas and Osaka Gas have now also started to market fuel cell-based systems. In order to field test new fuel cell-based micro-CHP systems on a large scale and prepare their market entry, the two utilities have introduced leasing schemes in which residential customers can test fuel cell-based micro-CHP systems in their homes (IEA, 2008b).

Volkswagen/LichtBlick (Germany)

In Germany, car manufacturer Volkswagen is partnering with the energy company LichtBlick in the production and distribution of natural gas powered small CHP systems. Volkswagen produces the CHP systems, which run on the same engines as used in Volkswagen's Touran and Caddy automobiles. The 20 kW_e systems are equipped with a heat storage, grid connection and remote data monitoring devices. LichtBlick distributes the systems by an approach similar to leasing the equipment to customers: the company rents the client's 'boiler-room', and is responsible for installation and service, maintenance and repairs of the equipment. LichtBlick also assumes responsibility for dismantling the customer's old gas heating system. The building owner or user pays an initial contribution of €5000 for the installation, which is significantly cheaper than the installation of a new heating system. In addition, the building owner or user pays for the heat he consumes and a flat rate for maintenance, but receives a monthly 'rent' of €5 for the boiler room and 0.5 €ct from the German CHP feed-in tariff for every kWh of power fed back into the grid.

According to the company, the CHP system can reduce energy consumption by up to 30–40 per cent compared with conventional heat and power supply. For the scheme to be financially attractive, customers need to have a heat demand of at least 40,000 kilowatt hours. This means that the CHP plants are suitable for very large single-family homes, for buildings with two or three flats, small businesses or hotels/B&Bs and public and social facilities such as schools and churches.

In November 2010, LichtBlick installed the first systems at residential and commercial customers in Hamburg, Germany. The company has been running a test set-up with 25 decentralised plants at Volkswagen's production facilities in Salzgitter since early 2010. In the long term, the two companies have ambitious targets, planning to generate a network of 100,000 of these home power plants. With these the company plans to 'create' a large, virtual (2 GW_e) dispatchable power plant. As LichtBlick is responsible for the operation of all of the plants, it would be able to profit from periods of high electricity prices by starting up the small CHP systems when the price exceeds a certain minimum level. In this way the thus created electric power could complement fluctuating power from renewable energy sources such as wind

Figure A.8 Lichtblick CHP system (Photo: Volkswagen)

and solar PV. The generated heat can be stored in the storage tanks to be extracted by the building user when needed. In addition, LichtBlick is able to profit from the German CHP feed-in tariffs. The combination of these aspects creates a viable business model.

(Based on Volkswagen, 2010, 2011)

A.8 Leasing of a heat pump system

(Business model: leasing of RET; energy supply contracting; see pages 87, 88, 108)

The following presents the example of a lease of a large heat pump system combined with heat-and-cold storage in a large (35,000 m²), new office building in the Netherlands. The building is currently being used as national headquarters for the company T-Mobile which rents the building. The heat pump system supplies heating and cooling to the building.

The lease was originated and organised in the following way:

- A real estate developer (the initial building owner) realised the construction of the building including the renewable energy system.
- For engineering and installation of the system, the developer hired an external energy consultancy and an energy service provider.

- After the system was installed, it was 'outsourced' to the energy service provider, who took over legal and economic ownership of the system from the building owner.[3] The energy service provider in turn agreed to an energy supply contract for 15 years with the building owner (who thus outsources the energy supply), and maintains the system.
- After the building was finalised it was also sold to an investor, who is renting the building to a commercial organisation (T-Mobile) as tenant.

The following describes the financing structure of the lease:

- The system was fully financed via a financial lease. The energy service provider invested in the system and leases it to the building owner. However, to reduce risk, the energy service provider charged the majority (80 to 90 per cent) of the total investment of €773,000 to the building owner at once. The energy service provider charges the remaining investment costs to the tenant spread over 15 years via the rent charged for the building.
- The energy service provider also charges the building owner yearly maintenance costs of €24,000 for the system. The building owner in turn recovers these maintenance costs by periodically charging the tenant a fixed service fee.
- Overall, the tenants' payments for renting the building including the heat pump system are based upon a share of investment costs, maintenance costs, and energy supply costs.
- The costs for heating and cooling (not electricity and lighting) for the tenant are expected to be 10 to 20 per cent lower than with a conventional energy system. Note that in addition to energy costs, the tenant pays the fixed service fee for maintenance and a higher rent to recover the initial investment costs of the energy service provider. The maintenance costs are expected to be lower than for a conventional system.
- Because of the economic ownership and as a commercial party, the energy service provider was allowed to receive a tax benefit on the investment. Half of this tax benefit was passed through to the tenant via a reduction in the rental price.
- The building owner is expected to have the largest cost savings after 15 years when renovation investments are required. Since a heat pump system is cheaper to replace than a conventional system, total cost savings for the owner are expected to be €210,000.

(Based on AgentschapNL, 2010)

A.9 Energy supplier obligations in the UK

(Business model: energy saving obligation; see pages 95–6)

The following illustrates how an energy saving obligation is implemented in practice and how it can lead to the development of business models.

Background

Energy supplier obligations in the UK are enforced by the policy programme 'Carbon Emission Reduction Target (CERT)'. The CERT programme obliges all large domestic energy suppliers to realise energy savings in households. By doing so, energy suppliers are required to deliver measures that will provide overall lifetime carbon dioxide savings of 293 $MtCO_2$ by December 2012. CERT is implemented in several phases; in 2011 the fourth phase started. Energy saving obligation schemes began in 2002 in the UK under a programme called 'Energy Efficiency Commitment' (EEC) programme which was the predecessor to CERT and ran until 2008. As of 2013, the obligations will be enforced by a new policy programme, the Energy Company Obligation (ECO) as established in the UK Green Deal.

CERT has clear energy demand and emission reduction targets with indirect aims such as reducing fuel poverty, securing jobs and realising social benefits like improved air quality and comfort. The programme targets 'priority groups', namely low income households and elderly private home owners of age 70 and older.

Structure of the programme

The basic business model for the energy suppliers is that they are obliged to invest in energy saving measures to be installed at their customers with the aim of reaching an energy savings target. In return, energy suppliers are allowed to pass on their investment costs by increasing the energy prices they charge to all their customers. This structure is shown in Figure A.9.

Stakeholders

Key stakeholders of the programme are the Department of Energy & Climate Change (DECC) as initiator, Ofgem (the energy market regulator) who administrates and monitors the programme, six large energy suppliers who have the obligation, households where savings are realised, and installers. The basic assumption underlying the programme is that energy suppliers working under such market mechanisms will realise the carbon reduction target more efficiently than the government would through a centralised programme.

The energy suppliers hire installers to deliver measures. Furthermore, energy suppliers have partnerships with several parties (housing corporations, municipalities, manufacturers of energy efficient products, etc.) who also promote and realise measures at the target groups. Energy suppliers will get a penalty from Ofgem if they do not reach their targets.

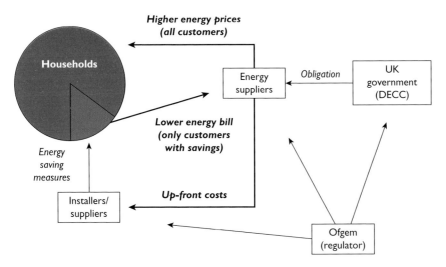

Figure A.9 Schematic representation of key actors in the UK CERT programme

Energy suppliers pay the majority of the investment costs, fully or partly, of the saving measures either to home owners or to suppliers like manufacturers and retailers. Investment costs are estimated to be on average £1 billion (around €1,16 billion)[4] annually (Ofgem, 2010). Households do not finance any costs themselves, as average investment costs per households are estimated only at £500 (around €590). The largest financial risk of the scheme thus lies with the energy suppliers who need to earn their investments back via higher energy prices spread over a number of years. This level of investment is unlikely to be sufficient in the future, as only the more expensive saving potential remains. The future ECO system will therefore require households to finance part of the full investment themselves via an on-bill financing scheme.

Results achieved

In the current phase of the programme, a large part of the energy savings are delivered by improving building insulation, mainly through cavity and loft insulation which are the most cost-efficient eligible measures (see Figure A.10). Efficient lighting, heating (heat pumps) and appliances are also common measures. The current programme restricts certain measures, e.g. the use of energy saving lamps (CFLs) which have already been massively supported by the previous UK energy supplier obligation schemes.

It is estimated that over the lifetime of measures implemented between April 2008 and December 2011, the CERT programme realises around 181 Mt CO_2 emission reductions (Ofgem, 2011). With these savings, the programme is expected to be on track to achieve the targeted results. The

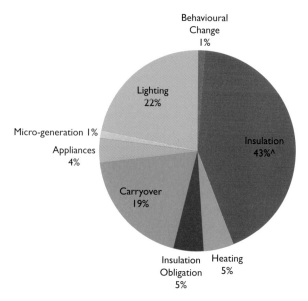

Figure A.10 Carbon savings by measure type in the third year of the Carbon Emissions Reduction Target (CERT) (Source: Ofgem, 2011)

government expects social benefits to households, including energy saved, at around £22 billion (around €26 billion) over the period 2008–2012 (DECC, 2011b).

Conclusion

The main strength of the programme is the obligation to energy suppliers, which addresses many barriers to energy savings, such as low priority of energy investments, lack of up-front capital for investments etc., and leads to significant savings compared to voluntary programmes. The CERT is very attractive to participating home owners, as energy saving measures are financed by energy suppliers and installation is taken care of. The programme is able to realise savings in the existing housing stock, where it is generally difficult to realise energy savings through voluntary measures or purely commercial activities. The focus on low income households reduces the chances of free riding, as this target group is less likely to have taken the measures anyway. However, supplying certain measures for free or at a low price can reduce the effectiveness of the programme. An example are CFLs which were actually oversupplied and partly ended up unused. The consequences are unnecessary supplier costs and a limited energy savings impact.

(Based on CERT, 2011; DECC, 2010, 2011a, 2011b, 2012a, 2012b; Eceee, 2011; Ofgem, 2010, 2011)

A.10 Power Smart Residential Loan programme of Manitoba Hydro, Canada

(Business model: on-bill financing; see pages 80–1)

The on-bill financing programme of Manitoba Hydro, called Power Smart Residential Loan, is the most successful on-bill financing programme in Canada and, measured by its loan volumes and loan values, the largest in North America. The programme started in 2001 and is ongoing. Since the beginning of the loan programme, Manitoba Hydro has disbursed more than $200 million through over 50,000 loans. As the name implies, the programme focuses on the residential sector which is considered more stable than the market for small businesses. As of March 2012, the minimum loan size was $500 and the maximum size $7,500 (approximately between €370 and €5,500). The maximum term is 5 years at 3.9 per cent interest (as of March 2012). If the loan (or part of the loan) is for the purchase of a high efficiency natural gas furnace, the loan has a maximum term of 15 years. The interest rates have been adjusted various times over the lifespan of the programme and are currently at a comparatively very low level. Default levels have generally been lower than 1 per cent. The programme is administered by the utility; the execution is undertaken by a strong network of contractors. For the customer, the loan repayments are added as a line item to the utility bill. At the sale of the house, the loan becomes due and needs to be paid back.

The following measures are eligible for financing under the programme: adding insulation, installing ventilation, sealing air leaks, replacing windows and doors, upgrading the existing natural gas or electric heating system, and domestic water heaters. Geothermal heat pumps are financed through another Manitoba Hydro offering, the Earth Power Loan, which finances a maximum of $20,000 (about €14,700) over 15 years. When looking at the period from March 2001 to 2008/2009, $167 million in loans were disbursed: 59 per cent of these loans were used to finance energy efficient window and door upgrades, 35 per cent for heating system upgrades and 6 per cent for combination of upgrades to insulation, ventilation and air sealing. In addition, Manitoba Hydro offers rebates for some measures, e.g. for insulation measures. These rebates partly explain the fact that insulation measures are not financed that much through the Power Smart Residential Loans, as only the balance between the total cost of measures and the rebates are financed through the loan. The Power Smart Residential Loan is intended as a cost recovery programme by Manitoba Hydro where the administration costs are recovered through the interest rate charged to the client. At times, Manitoba Hydro has subsidised the interest rate from its profits depending on the direction and focus of the energy efficiency and greenhouse gas emissions reduction targets set by the provincial government.

Reasons for the success of the programme are considered to be the fact that the loan application process is streamlined and that Manitoba Hydro has strong relationships with the contractors and retailers who execute the programme. Loan approvals normally are processed within one business day. Payments to the contractor happen within 20 business days. All suppliers have received training on how the programme procedures work and have signed a supplier participation agreement.

(Based on Manitoba Hydro, 2011, 2012;
Fuller, 2008; Brown & Conover, 2009)

A.11 LEED certification

(Business model: green building certification; see p. 56)

The 'Leadership in Energy and Environmental Design' (LEED) standard, administered by the US Green Building Council (USGBC), is one of the largest voluntary green building standards globally. A building can acquire one of the four LEED certification levels: certified, silver, gold and platinum with platinum being the highest rating. Criteria for achieving these ratings vary, e.g. between new and existing buildings and different building types such as residential buildings, commercial buildings and neighbourhood developments.[5]

According to the US Green Building Council's public database,[6] there are 45,000 registered and LEED certified building projects worldwide up to 2012. The database shows that more than half of the buildings are privately owned. The remaining buildings are owned by the public sector (institutional, for example schools, or government owned). The average size of the buildings in this database is around 200,000 gross m², although most of the buildings are smaller. Almost 90 per cent of all LEED projects are realised in the US. Other countries with LEED certified developments include the United Arabic Emirates, Brazil, Canada, China and India.

The following illustrates three case studies where LEED certification was achieved for new buildings.

Pearl Place, Maine, US

Pearl Place is a workforce housing project which was completed at the end of 2007. It consists of 60 affordable apartments comprising of a mix of one-, two- and three-bedroom apartments. The project is owned and was developed by Avesta Housing, which is Maine's largest non-profit affordable housing developer. The buildings in the Pearl Place development were certified LEED Gold in 2008. Energy efficient features of the buildings include super-insulation and a tight building envelope, unit compartmentalisation, energy efficient fixtures, appliances and mechanical equipment, low-Volatile Organic Compound (VOC) finishes and air-to-air heat exchange. Non-energy related green aspects include the use of high quality, permanent building

Figure A.11 Pearl Place exterior
(Photo courtesy of Avesta Housing)

materials such as brick and fibreglass and recycling facilities. Reasons for pursuing the LEED certification were related to municipal regulations which require building developments that receive funds from the City of Portland to achieve at least LEED Silver certification. However, according to one of the involved stakeholders one of the greatest benefits gained from the integrated process used was that the development cost approximately the same as other multi-family affordable housing projects in the region. Thus, other developers realised that green buildings do not have to come at an additional cost to most building owners. However, the costs for the certification and required consultants may still pose a challenge to project developers.[7]

Moffett Towers, California, US

The Moffett Towers complex is a new office and R&D campus in the centre of California's Silicon Valley. Lot 1 of the development consists of three office towers and two parking garages, totaling 866,000 square feet, in the heart of the complex. Lot 1 achieved LEED for Core & Shell Gold certification in 2009.

In terms of energy performance, Lot 1 is expected to use 30 per cent less energy than a building which follows the baseline building code. This is achieved through a high efficiency heating and air conditioning system with variable speed fans, a sophisticated project automation system and a high efficiency lighting design. The fitness centre located on the premises uses a solar thermal hot water system for pool heating, which constitutes about 10 per cent of the building's overall energy usage. However, solar PV systems were considered to be too expensive. The design process was supported by energy modeling which, for example, demonstrated that due to the site constraints it was not possible to construct all of the buildings according

Figure A.12 Insulates glazing and sunshades of Moffet Towers building (Photo courtesy of DES Architects and Engineers)

to an ideal north–south orientation. Moreover, the modeling provided criteria which supported the selection of the most efficient glazing to use and of the right amount of insulation required to meet the LEED criteria. Other important sustainability related features of the project are related to water consumption: the project is designed to use 40 per cent less water than a building that has conventional fixtures installed.

Suzlon One Earth, India

Suzlon, one of the world's largest wind turbine manufacturers, developed a new office complex for its global headquarters in India, the Suzlon One Earth office complex, which is considered to be among the greenest office buildings in India. It was certified according to LEED for New Construction Platinum certification from the India Green Building Council in 2010. Five per cent of its annual energy requirements is generated on-site through conventional and building-integrated PV panels and on-site wind turbines. The remaining electricity requirements are generated through Suzlon's off-site wind turbines Thus the office complex is technically a zero-energy project. Moreover, energy savings are achieved through, for example, the use of LED lighting systems, solar water heating and evaporative cooling. The operating expenses for the evaporative cooling system are more than 20 per cent lower than for a standard air conditioning system and the new system also had lower capital costs. Overall, the buildings' energy performance per square metre of office space reflects energy savings of more than 45 per cent over

Figure A.13 Suzlon One Earth (Photo by Sunil Rikhaye)

conventional office buildings in India. Annual energy audits show that, so far, the energy performance is as planned. With the LEED certification of the project, Suzlon aims to reflect its goal of promoting clean power globally. And according to the LEED Project Administrator, the project development process was less costly than for commercial structures of a comparable size, and led to lower post-commissioning operating costs.

(Based on USGBC, 2011a, 2011b, 2011c, 2011d)

A.12 Berlin Energy Saving Partnership

(Business model: energy performance contracting; see pages 35–38)

The Berlin Energy Saving Partnership was jointly developed by the Berlin Energy Agency and the Berlin's Senate Department for Urban Development in 1996. It is a model for achieving energy savings through Energy Performance Contracting (EPC), tapping into the potential for energy savings in a pool of public buildings with different properties. Since 1996, within the Berlin Energy Saving Partnership, 26 energy partnerships were launched, comprising more than 500 public buildings. Some of the earlier contracts have already expired by now; thus as of May 2011, about 375 building were under EPC contracts. The project is ongoing and the model has also been replicated in other regions of Germany. The latest building pool in Berlin was contracted in mid-2011. Examples of public buildings upgraded in the frame of the project in Berlin include town halls, schools, day nurseries etc.

In the set-up of the Berlin Energy Saving Partnership, the Berlin Energy Agency acts as the independent market and project facilitator, who moderates and manages the process, e.g. the negotiations on the contract, and puts the building pools out for bidding. The EPC contracts are implemented by private ESCOs (energy service companies) which finance investments into energy savings. The ESCOs undertake the up-front investment into energy saving measures and recover these initial costs through energy cost savings over the contract period, which is on average around 12 years. Average payback periods of the investments undertaken are about 5 years. Typical energy saving measures applied are efficient lighting, heating control systems, and energy consumption regulators; occasionally, insulation and CHP systems are applied as well.

The ESCO is also responsible for the planning, implementation and management of the energy savings measures and bears all the operational and economic risk of the project over the entire project term. The contractor bears the responsibility for the operational performance of the technical systems, including any risks caused by a breakdown of the systems. The ESCO legally guarantees a minimum level of energy savings. This implies that if the targeted energy savings are not achieved, the ESCO will still compensate the building owner for them (Berliner Energieagentur, 2006). Additional cost savings are shared by the ESCO and the building owner, which is an additional incentive for the ESCO and the building owner to participate. Once the contract period ends, the full energy cost savings accrue to the building owner.

In the frame of the project, public buildings, typically from one administration, are 'pooled' to reduce transaction costs. This also makes it possible to include less profitable buildings in the pool. Building pools that participate in the Berlin Energy Saving Partnership must have a minimum annual energy bill of approximately €200,000. The average energy cost baseline is about €1.8 million/a. The number of buildings per pool varies: some contracts include only one building, e.g. a hospital with significant energy use on its own; the largest building pool comprises of 73 buildings.

For the building owners, the advantage of the model is that they do not bear any investment costs, can outsource the implementation of the energy saving measures as well as the technical and economic risks, and realise energy cost savings.

The local government in Berlin subsidises the services carried out by the Berlin Energy Agency by 50 per cent (New York City Global Partners, 2011). This support is critical as otherwise most building owners would not be willing to engage in the EPC project (Berliner Energieagentur, 2007).

In the 26 building pools contracted under the Energy Saving Partnership overall about €11.3 m of guaranteed annual costs savings are achieved,

Figure A.14 Applying insulation at the facade of a building at Kornburger Weg in Berlin (Picture: Berliner Energieagentur GmbH – www.berliner-e-agentur.de)

Figure A.15 View into the inside of a decentralised CHP plant (Picture: Berliner Energieagentur GmbH – www.berliner-e-agentur.de)

€2.7 m of which are costs savings for the government of Berlin (Berger, 2011; Berliner Energiagentur, 2011). This is accomplished through overall contractually guaranteed minimum investments of €51.6 m. An example building pool are 69 schools, kindergartens and gyms in Berlin's district Steglitz-Zehlendorf (Berger, 2011). These 69 buildings have an energy cost baseline of €1.84 million/a. The performance contract for the building pool foresees guaranteed savings of 29.4 per cent or €541,679 /a, achieved through an investment of about €2.8 m. The contract has a duration of 14 years and includes the following measures: new boilers in 11 buildings, a switch from coal/heating oil to gas, building automation, the modernisation of lighting systems and investment of €100,000 into renewable energy technologies such as solar thermal systems. These measures lead to an expected CO_2 reduction of 3,973 t/a (Berger, 2011).

Energy performance contracting as undertaken by the Berlin Energy Saving Partnership is a well replicable concept which can lead to significant energy cost savings in public buildings without the need for up-front capital investments by the involved public building owners. However, it does require independent facilitators to develop and facilitate projects, a functioning market of ESCOs which have sufficient access to capital to bear the significant up-front investment costs. And the concept normally does not extend to measures targeting the building shell or renewable energy carriers. Similar initiatives are being implemented outside of Berlin in other German regions, but also in Bulgaria, Slovenia, Romania and Chile. Moreover, know-how has been transferred to help initiate similar programmes in Central, Eastern and Western Europe.

(Based on Berger, 2011; Berliner Energieagentur, 2006, 2007, 2011; New York City Global Partners, 2011)

Notes

1 See for example Nuon (www.nuon.nl) and Eneco (www.eneco.nl) (websites consulted 26 October 2011).
2 A low energy performance (i.e. high energy prices for the tenant) may lead to a

freeze of the rental price. This is stipulated by a temporary provision of law (Eerste Kamer, 2011).

3 As the building is owned by the investor, to get ownership of only the energy installations the energy service provider had to obtain building and planting rights for the installations via a notarial act.

4 Calculations in this case study are based on the average exchange rate Euro to Pound in 2010.

5 Criteria for achieving LEED certification fall, for example, into the following categories: Sustainable Sites, Water Efficiency, Energy and Atmosphere, Materials and Resources, Indoor Environmental Quality and Innovation in Design. Depending on what is being certified, the categories vary, e.g. for LEED for Neighbourhood Development categories such as Neighbourhood Pattern and Design, and Green Infrastructure and Buildings are used.

6 See the website http://www.usgbc.org/LEED/Project/CertifiedProjectList.aspx, consulted 13 February 2012.

7 Whilst the cost for a combined design and construction review of a new building of 4,650 m^2 or less applying for LEED certification is around USD 2,500, the costs for consultants advising during the project development process on how to achieve certification can be significantly higher.

References

Activum finance (2011) Available at http://www.activum-finance.nl/informatie. Last accessed 24 June 2011.

Aedes (2009) Woonlastenwaarborg bij energiebesparing. May 2009. Available at http://www.bespaarenergiemetdewoonbond.nl/art/uploads/Folder%20Woonlasten waarborg%20Aedes%20en%20Woonbond_1246970996.pdf. Last accessed 23 September 2012.

Aedes (2011) *Energielabels in woningwaarderingsstelsel*. Available at: http://www.aedesnet.nl/content/artikelen/achtergrond/unknown/dossier-huur-en-verhuur/Energielabels-in-woningwaarderingsstelsel.xml. Last accessed 23 September 2012.

AgentschapNL (2010) *Dure plannen, goedkope oplossingen – Adviseren over organisatie en financiering van energiebesparing in de utiliteitsbouw*. SenterNovem, Kompas energiebewust wonen en werken.

AgentschapNL (2011) *Stimuleringsregeling Duurzame Energie*. Available at http://www.agentschapnl.nl/nl/programmas-regelingen/stimulering-duurzame-energieproductie-sde. Last accessed 30 June 2011.

Bailey, M. & Broido Johnson, C. (2009) *The American Recovery and Reinvestment Act (ARRA) of 2009: innovative energy efficiency financing approaches*. US Department of Energy, Office of Energy Efficiency and Renewable Energy.

Berger, S. (2011) *Energy Saving Partnership – Better Practice Berlin*. Berliner Energieagentur. Presentation at the BETTER-PRACTICE-EXCHANGE 2011 – ENERGIE, Potsdam, 30 May 2011. Available at http://ecologic.eu/files/attachments/Projects/2358/prasentation_berger.pdf. Last accessed 6 September 2012.

Berliner Energieagentur (2006) *Performance contracting – Energy Saving Partnership – A Berlin success model*. Available at www.dfhk.fi/fileadmin/ahk.../Berlin_Vortrag_BEA_englisch.pdf. Last accessed 6 September 2012.

Berliner Energieagentur (2007) *Case study Energy Savings Partnership*. C40 Large Cities Climate Summit, New York City.

Berliner Energiagentur (2011) *Übersicht der Energiesparpartnerschaften in Berlin – Stand Mai 2011*. Available at http://www.berliner-e-agentur.de/sites/default/files/uploads/pdf/pooluebersichtespaktuell.pdf. Last accessed 6 September 2012.

Bertoldi, P. & Rezessy, S. (2009) *Energy saving obligations and tradable white certificate schemes*. CEC/JRC report. Ispra, Italy.

Bertoldi, P., Boza-Kiss, B. & Rezessy, S. (2007) *Latest development of energy service companies across Europe – a European ESCO update*. EC JRC Institute for Environment and Sustainability, Ispra, Italy.

Bertoldi, P., Rezessy, S., Lees, E., Baudry, P., Jeandel, A. & Labanca, N. (2009) Energy supplier obligations and white certificate schemes: comparative analysis of experiences in the European Union. *Energy Policy* 38, 1455–1469.

Bleyl, J. & M. Suer (2006) *Comparison of different finance options for energy services.* In light+building, International Trade Fair for Architecture and Technology, Frankfurt, 2006.

Bleyl, J. W. (2009) *Integrated energy contracting (IEC). A new model to combine energy efficiency and (renewable) energy supply.* IEA DSM Task XVI Discussion Paper. Available at www.ieadsm.org. Last accessed 2 February 2011.

Bleyl, J. W. (2011) *Conservation first! The new integrated energy-contracting model to combine energy efficiency and renewable supply in large buildings and industry.* In ECEEE Summer Studies, paper ID 485, Belambra Presqu'île de Giens, France, June 2011.

Bleyl, J. W. & Schinnerl, D. (2008a) 'Energy contracting' to achieve energy efficiency and renewables using comprehensive refurbishment of buildings as an example. In *Urban Energy Transition,* ed. Peter Droege, Elsevier. Available at www.ieadsm.org

Bleyl, J. W. & Schinnerl, D. (2008b) *Opportunity cost tool.* Preliminary version. Graz Energy Agency, May 2008.

BMU (2011) *Richtlinien zur Förderung von Maßnahmen zur Nutzung erneuerbarer Energien im Wärmemarkt* Vom 11. März 2011, Berlin, March 2011.

BMWFJ (2012) *Bundescontracting.* Bunderministerium für Wirtschaft, Familie, Jugend website. Available at http://www.bmwfj.gv.at/Tourismus/energieeinsparungen/Seiten/Bundescontracting.aspx. Last accessed on 24 September 2012.

Boland, C. (2010) *Local governments and federal agency clash over property assessed clean energy programs.* DEFG's Series of Regulatory Choices, No. 3, September 2010.

Boonekamp, P. & Vethman, P. (2010) *Analysis of policy mix and development of Energy Efficiency Service.* Task 2.3 of the project 'Change Best – Promoting the development of an energy efficiency service (EES) market'.

Boot, P. A. (2009) *Energy efficiency obligations in the Netherlands – a role for white certificates?* ECN, September 2009.

Brealey, R. A. & Myers, S. C. (2003) *Principles of corporate finance.* Mcgraw-Hill/Irwin Series in Finance, Insurance & Real Estate, Mcgraw Hill Higher Education.

Brown, M. H. & Conover, B. (2009) *Recent innovations in financing for clean energy.* Southwest Energy Efficiency Project, Boulder, Colorado, October 2009.

Brown, M. (2009) *Paying for energy upgrades through utility bills.* State Energy Efficiency Policies: Options and Lessons Learned. Brief # 3. Alliance to Save Energy.

Brown, M., Chandler, J. & Lapsa, M. (2010) *Policy options targeting decision levers: an approach for shrinking the residential efficiency gap.* 2010 ACEEE Summer Study on Energy Efficiency in Buildings.

Bunse, M., Irrek, W., Siraki, K. & Renner, G. (2010) *National report on the energy efficiency service business in Germany.* Task 2.1 of the project 'Change Best – Promoting the development of an energy efficiency service (EES) market'.

CalCEF (2009) *New business models for energy efficiency.* CalCEF's Innovation White Paper, March 2009.

CERT (2011) *Carbon Emission Reduction Target policy programme UK.* Available at http://www.decc.gov.uk/en/content/cms/funding/funding_ops/cert/cert.aspx. Consulted on 9 March 2011.

City of Berkeley (2010) *Berkeley FIRST final evaluation*. City of Berkeley, Office of Energy and Sustainable Development, November 2010.

Clearsupport (2008) *Recommendations on financial schemes for local Clearinghouse operation*. Final draft version, Clearsupport project. Berliner Energie Agentur, Investitionsbank Schleswig-Holstein & Clearsupport project partners, July 2008.

COWI (2008) *Promoting innovative business models with environmental benefits*. Final report to the European Commission.

DECC (2010) *Extending the Carbon Emissions Reduction Target to December 2012. Impact assessment*. UK Department of Energy & Climate Change, June 2010.

DECC (2011a) Personal communication Mr Alan Clifford of DECC, by mail, March 2011.

DECC (2011b) *Renewable heat incentive*. UK Department of Energy and Climate Change, April 2011.

DECC (2011c) *GB Energy company obligations (ECO)*. Available at http://re.jrc.ec.europa.eu/energyefficiency/events/pdf/JRC%20WhC%20workshop/session%201/ITALY%20PRESENTATION.pdf. Alan Clifford, Department of Energy and Climate Change. Eceee workshop on energy saving obligations, September 2011.

DECC (2012a) Personal communication Mr Alan Clifford of DECC, by mail, 11 January and 7 February 2012.

DECC (2012b) *Energy consumption in the United Kingdom*. Available at http://www.decc.gov.uk/en/content/cms/statistics/publications/ecuk/ecuk.aspx. Consulted 12 January 2012.

Dsireusa (2011) *Overview of renewable energy policy in the US*, website: http://www.dsireusa.org/summarytables/rrpre.cfm. Consulted 21 July 2011.

EACI (2011) Boosting the Energy Services Market in Europe – Conclusions – IEE Workshop, Brussels, 23 February 2011. Executive Agency for Competitiveness and Innovation (EACI).

ECN (2011a) Personal communication from C. Tigchelaar, 24 June 2011.

ECN (2011b) *Energy efficiency directive impact assessment report*. ECN, June 2011.

Ecorys, Ecofys and BioIntelligence (2010) *Study to support the impact assessment for the EU Energy Saving Action Plan*.

Eceee (2011) *Workshop on energy saving obligations*, September 2011.

Eerste Kamer (2011) *Wijziging van de Uitvoeringswet huurprijzen woonruimte (wettelijke grondslag verschillende waardering energieprestaties huurwoningen*. Ministerie van BZK, Eerste Kamer, vergaderjaar 2010–2011, 32 302, B.

Eichholtz, P., Kok, N. & Quigley, J. M. (2008) *Doing well by doing good?* Institute of Business and Economic Research, Berkeley Program on Housing and Urban Policy, University of California, Berkeley.

Eikmeier, B., Seefeldt, F., Bleyl, J.W. & Arzt, C. (2009) Contracting im Mietwohnungsbau. In *Schriftenreihe Forschungen Heft 141*, Bundesministerium für Verkehr, Bau- und Stadtentwicklung, Bonn, April 2009.

ESP Berlin (2009) *Energiesparpartnerschaft Berlin. Ergebnisse aus 23 Gebäudepools*. Berliner Energieagentur, Berlin, unpublished.

Essent (2011) Personal communication Mr G. Kempen, July 2011.

European Commission (2002) Directive 2002/91/EC of the European Parliament and of the Council of 16 December 2002 on the energy performance of buildings.

European Commission (2006) Directive 2006/32/EC of the European Parliament and of the Council on Energy End Use and Energy Services, as of 5 April 2006.

European Commission (2008) *Energy sources, production costs and performance of technologies for power generation, heating and transport.* Commission Staff Working Document accompanying the Communication from the Commission to the European Parliament, the Council, the European Economic and Social Committee and the Committee of the Regions, Second Strategic Energy Review, An EU Energy Security and Solidarity Action Plan, SEC(2008)2872.

European Commission (EC) (2009a) *Impact assessment guidelines.* Available at http://ec.europa.eu/governance/impact/commission_guidelines/docs/iag_2009_en. pdf. Last accessed 26 September 2012.

European Commission (2009b) *Low energy buildings in Europe: current state of play, definitions and best-practice.* Information note, Brussels, 25 September 2009.

European Commission (2010/11) *Concerted Action – Energy Services Directive.* See http://www.esd-ca.eu/CA-ESD

European Commission (2010) Directive 2010/31/EU of the European Parliament and of the Council of 19 May 2010 on the energy performance of buildings (recast).

European Committee for Standardization (EN) (2009) *Energy efficiency services – definitions and essential requirements.* CEN/CLC/TF 189, draft under discussion, March 2009.

European Parliament (2012) *Compromise amendments on the proposal for a directive of the European Parliament and of the Council on energy efficiency.* Draft report, 2011/0172(COD), 22.02.2012.

Fina-ret (2008) *Fina-ret project – WP 2,* review of existing financing products/ mechanisms for RE and EE Technologies Applications, May 2008.

Franklin Energy (2011) *Financing energy improvements – insights on best practices to engage stakeholders and marry dollars with demand.* Office of Energy Security, Minnesota Department of Commerce, January 2011.

Fresh project (2011) *Energy retrofitting of social housing through energy performance contracts.* Fresh project, January 2011.

Foxona, T. J., Grossa, R., Chaseb, A., Howesb, J., Arnallc, A. & Anderson, D. (2005) UK innovation systems for new and renewable energy technologies: drivers, barriers and systems failures. *Energy Policy*, 33: 2123–2137

Fuerst, F. & McAllister, P. (2008) *Does it pay to be green? Connecting economic and environmental performance in commercial real estate markets.* The University of Reading, Business School.

Fuerst, F. & McAllister, P. (2011) Green noise or green value? Measuring the effects of environmental certification on office values. *Real Estate Economics*, 39(1): 45–69.

Fuller, M. (2008) *Enabling investments in energy efficiency – a study of energy efficiency programs that reduce first-cost barriers in the residential sector.* California Institute for Energy and Environment, September 2008.

Gifford, J. S., Grace, R. C. & Rickerson, W. H. (2011) *Renewable energy cost modeling: a toolkit for establishing cost-based incentives in the United States.* NREL, May 2011.

Gray, S. J. & Needles, B. E. (1999) *Financial accounting – a global approach.* South-Western College Pub.

Greenchoice (2011) *Zonvast.* Available at http://www.greenchoice.nl/thuis/zelf-opwekken/met-zon/ZonVast. Last accessed 11 July 2011.

IEA (2007) *Mind the gap – quantifying principal agent problems in energy efficiency.* Available at http://www.iea.org/textbase/nppdf/free/2007/mind_the_gap.pdf. Last accessed 27 June, 2011

IEA (2008a) *Promoting energy efficiency investments – case studies in the residential sector.* Available at http://www.iea.org/textbase/nppdf/free/2008/Promoting EE2008.pdf. Last accessed 27 June 2011.

IEA (2008b) *CHP/DHC country scorecard: Japan.* The International CHP/DHC Collaborative – Advancing Near Term Low Carbon Technologies.

IEA (2010) *Money matters – mitigating risk to spark private investments in energy efficiency.* Philippine de T'Serclaes. Information Paper, available at http://www.iea.org/papers/efficiency/money_matters.pdf, September 2010. Available at http://www.iea.org/papers/efficiency/money_matters.pdf. Last accessed 27 June 2011.

IEA-RETD (2007) *Renewables for heating and cooling – untapped potential.* Available at http://www.dekoepel.org/documenten/Renewable_Heating_Cooling_Final_WEB.pdf. Last accessed 13 September 2011.

IEA-RETD (2010) *Best practices in the deployment of renewable energy for heating and cooling in the residential sector.* Available at http://www.iea-retd.org/files/IREHC%20Final%20Report%2020100726.pdf. Last accessed 25 November 2011.

IEE workshop (2011) *Boosting the energy service market in Europe – conclusions.* IEE workshop, February, 2011.

Institute for Building Efficiency (2010a) *Property assessed clean energy financing – benefits and barrier busting.* Available at http://www.institutebe.com/InstituteBE/media/Library/Resources/Financing%20Clean%20Energy/PACE-Benefits-and-Barrier-Busting.pdf. Last accessed 26 September 2012.

Institute for Building Efficiency (2010b) *PACE: opportunities for owners through tax lien financing.* Available at http://www.institutebe.com/clean-energy-finance/pace-finance.aspx. Last accessed 11 May 2011.

Institute for Building Efficiency (2010c) *Unlocking the building retrofit market: commercial PACE financing – a guide for policy makers.* Available at http://www.institutebe.com/clean-energy-finance/pace-finance.aspx. Last accessed 11 May 2011.

IPCC (2007) Residential and commercial buildings. In *Climate Change 2007: Mitigation. Contribution of Working Group III to the Fourth Assessment Report of the Intergovernmental Panel on Climate Change.* B. Metz, O. R. Davidson, P. R. Bosch, R. Dave, L. A. Meyer (eds), Cambridge University Press, Cambridge, United Kingdom and New York, NY, USA.

IPCC (2011) Technical summary. In *IPCC special report on renewable energy sources and climate change mitigation.* O. Edenhofer, R. Pichs Madruga, Y. Sokona, K. Seyboth, P. Matschoss, S. Kadner, T. Zwickel, P. Eickemeier, G. Hansen, S. Schlömer & C. von Stechow (eds), Cambridge University Press, Cambridge, United Kingdom and New York, NY, USA.

IPMVP (2009) *Efficiency valuation organization (EVO) international performance measurement and verification protocol.* Available at www.evo-world.org/index.php. Last accessed 26 September 2012.

Johnson, K., Willoughby, G., Shimoda, W. & Volker, M. (2011) Lessons learned from the field: key strategies for implementing successful on-the-bill financing programs. *Energy Efficiency*, published online 13 January 2011.

JRC (2011) *JRC workshop on energy saving obligations*. Varese, Italy, January 2011. Available at http://re.jrc.ec.europa.eu/energyefficiency/events/WhC_Workshop.htm. Last accessed 19 August 2011.

Jumanjisolar (2011) Website, available at www.jumanjisolar.com/2011/06/autoconsumo-rentable-consumir-energia-fotovoltaica.html. Last accessed 26 September 2012.

Kats, G., James, M., Apfelbaum, S., Darden, T., Farr, D., Fox, R., Frank, L., Laitner, J., Leinberger, C., Saulson, G., Williams, S. & Braman, J. (2008) *Greening buildings and communities: costs and benefits*. Good Energies, Zug.

Labanca, N. (2010) *Status and development of the energy efficiency service business in 18 EU countries*. Task 2.1 of the project 'Change Best – Promoting the development of an energy efficiency service (EES) market'.

Langlois, P., S. Mansen (2012) World ESCO Outlook 2012, p.211f. Fairmount Press.

LBNL (2011) *Property Assessed Clean Energy (PACE) financing: update on commercial programs*. Policy Brief.

Lensink, S. M., van Tilburg, X., Mozaffarian, M. & Cleijne, J. W. (2007) *Feed-in support for renewable electricity – a comparison of three European implementations (Dutch)*, ECN-E-007-030, Amsterdam.

Leutgöb, K., Irrek, W., Tepp, J. & Coolen, J. (2011) *Strategic product development for the energy efficiency service market – the ChangeBest energy efficiency service development guide*. Available at http://www.changebest.eu. Last accessed 26 September 2012.

Limaye, D. R. (2011) *The emerging model of the public Super-ESCO: need and opportunity*. Presented at the Asia Clean Energy Forum, Manila, 23 June 2011.

Manitoba Hydro (2011) *Power Smart Residential Loan*. Available at http://www.hydro.mb.ca/your_home/residential_loan.shtml. Last accessed 12 March 2012.

Manitoba Hydro (2012) Personal communication by Colleen Kuruluk, Manager – Power Smart Marketing Programs, 27 March 2012.

Matthiessen, L. F. & Morris, P. (2004) *Costing green: a comprehensive cost database and budgeting methodology*. Davis Langdon. Available at http://www.davislangdon.com/USA/Research/ResearchFinder/2004-Costing-Green-A-Comprehensive-Cost-Database-and-Budgeting-Methodology/

McKinsey (2009) *Pathways to a low carbon economy – Version 2 of the Global Greenhouse Gas Abatement Cost Curve*. Available at www.mckinsey.com/global GHGcostcurve. Last accessed 30 September 2011.

Menichetti, E. & Touhami, M. (2007) *Creating a credit market for solar thermal: the PROSOL project in Tunisia*. Presented at the 3rd European Solar Thermal Energy Conference, 19–20 June 2007.

Milieucentraal (2011). *Website*. Available at http://www.milieucentraal.nl/pagina.aspx?onderwerp=zonneboiler#Zonneboiler_kopen:_subsidie. Last accessed 11 July 2011.

Miller, N., Spivey, J. & Florance, A. (2008) Does green pay off? *Journal of Sustainable Real Estate*. Available at http://www.costar.com/JOSRE/doesGreenPayOff.aspx. Last accessed 23 September 2012.

Missaoui, R. & Mourtada, A. (2010) *Instruments and financial mechanisms of energy efficiency measures in building sector*. WEC-ADEME case study on energy efficiency Measures and Policies. Available at http://www.worldenergy.org/documents/ee_case_study__financing.pdf. Last accessed 15 June 2011.

Nelson, A. (2008) *Globalization and global trends in green real estate.* RREEF Research, San Francisco, USA.

Nelson, A., Rakau. O. & Doerrenberg, P. (2010) *Green buildings – a niche becomes mainstream.* Deutsche Bank Research, 12 April 2010.

New York City Global Partners (2011) *Best practice: public-private partnership for building retrofits.* Available at www.nyc.gov/html/unccp/gprb/.../pdf/Berlin_Buildings_ESP.pdf. Last accessed 15 June 2011.

Nishizaki, Kunihiro (2008) *The Japanese experience in micro CHP for residential use.* Tokyo Gas, presented at Gas Industry Micro CHP Workshop, May 29, 2008.

Nostrand, J. M. van (2011) Legal issues in financing energy efficiency: creative solutions for funding the initial capital costs of investments in energy efficiency measures. *Journal of Energy and Environmental Law,* Winter 2011.

NREL (2010) *Property-Assessed Clean Energy (PACE) financing of renewables and efficiency.* Fact Sheet Series on Financing Renewable Energy Projects, National Renewable Energy Laboratory, Golden, Colorado.

Ofgem (2010) *A review of the second year of the Carbon Emissions Reduction Target – annual report to Secretary of State for Energy and Climate Change.* August 2010. Available at http://www.ofgem.gov.uk/Sustainability/Environment/EnergyEff/Documents1/CERT%20Annual%20report%20second%20year.pdf. Last accessed 23 September 2012.

Ofgem (2011) *Carbon Emissions Reduction Target – update.* Issue 14, December, 2011. Available at http://www.ofgem.gov.uk/Sustainability/Environment/EnergyEff/CU/Pages/CU.aspx. Last accessed 23 September 2012.

Osterwalder, A. (2004) *The business model ontology – a proposition in a design scheme approach.* PhD thesis at the University of Lausanne, Ecole des Haute Etudes Commerciale.

Osterwalder, A., Pigneur, P. & Tucci, C. L. (2005) Clarifying business models: origins, present and future of the concept, *CAIS (Communications of the Association for Information Systems),* 15, Article, May 2005.

OTB (2010) *Housing atatistics in the European Union 2010.* K. Dol & M. Haffner, OTB Delft, September 2010.

PACENow (2011) *PACENow website.* Available at http://pacenow.org/blog/. Last accessed 11 May 2011.

Patterson, W. (2010) *Toward real energy economics.* Presentation given at the International Association for Energy Economics Conference, Rio de Janeiro, 6–9 June 2010.

Pike research (2010) *Green building certification programs – global certification programs for new and existing buildings in the commercial and residential sectors: market analysis and forecasts.* Executive Summary, published Q2 2010.

Pivo, G. & Fisher, J. D. (2009) *Investment returns from responsible property investments: energy efficient, transit-oriented and urban regeneration office properties in the US from 1998–2008.* Working Paper, Responsible Property Investing Center, Boston College and University of Arizona, and Benecki Center for Real Estate Studies, Indiana University.

Porter, M. E. (2001) *Strategy and the Internet.* Harvard Business Review, Boston, March 2001.

Prognos (2009) *Prognos AG.* In *Contracting im Mietwohnungsbau.* Eikmeier, B., Seefeldt, F., Bleyl, J. & Arzt, C. Abschlussbericht, Bonn, April 2009.

RAEL (2009) *Guide to energy efficiency and renewable energy financing districts for local governments.* Prepared by Renewable and Appropriate Energy Laboratory (RAEL), University of California, Berkely for the City of Berkely, California.

Rappa, M. (2001) *Managing the digital enterprise – business models on the Web.* North Carolina State University.

REN21 (2011) *Renewables 2011 – global status report*, Paris, REN21 secretariat. Available from http://www.ren21.net/Portals/97/documents/GSR/REN21_GSR2011. pdf. Last accessed 25 October 2011.

Satchwell, A., Goldman, C., Larsen, P., Gilligan, D. & Singer, D. (2010) *A survey of the U.S. ESCO industry: market growth and development from 2008 to 2011.* Lawrence Berkeley National Laboratory, LBNL-3479E. June 2010. Available at http://www.naesco.org/resources/industry/documents/ESCO%20study.pdf. Last accessed 6 August 2011.

Schafer, B., Konstan, J. & Riedl, J. (2001) E-Commerce recommendation applications. *Journal of Data Mining and Knowledge Discovery* 5(1/2): 115–152.

Szomolanyiova, J. & Sochor, V. (2012) *Accelerating the development of the energy efficiency service markets in the EU – conclusions and policy recommendations from the ChangeBest project.* Available at http://www.changebest.eu/images/stories/ deliverables/changebest_policyconclusions_final.pdf. Last accessed 23 September 2012.

Tigchelaar, C., Daniels, B. & Menkveld, M. (2011) *Obligations in the existing housing stock: who pays the bill?* ECN, ECEEE summer study 2011. Available at http://www.ecn.nl/publicaties/PdfFetch.aspx?nr=ECN-M—11-070. Last accessed 19 September 2012.

UIPI (2010) *Revision of the Energy Efficiency Action Plan – strategy: contribution from private property and building owners' organisations to meet the challenge.* UIPI – GEFI – UEHHA.

UIPI & CEPI (2010) *Landlord/tenant dilemma.* Joint statement by CEPI, the European Council of Real Estate Professions and UIPI, the International Union of Property Owners, December 2010.

UK DECC (2010) *The Green Deal – a summary of the Government's proposals.* Available at http://www.decc.gov.uk/assets/decc/legislation/energybill/1010-green-deal-summary-proposals.pdf. Last accessed 10 July 2011.

UNEP (2007) *Assessment of policy instruments for reducing greenhouse gas emissions from buildings.* Report for the UNEP-Sustainable Buildings and Construction Initiative, available at http://www.unep.org/themes/consumption/pdf/SBCI_CEU_ Policy_Tool_Report.pdf. Last accessed 17 May 2011.

US Department of Energy (DOE) (2010a) *Guidelines for pilot PACE financing programs.* May 2010. Available at http://www1.eere.energy.gov/wip/pdfs/arra_ guidelines_for_pilot_pace_programs.pdf. Last accessed 26 September 2012.

US Department of Energy (DOE) (2010b) *Commercial Property Assessed Clean Energy (PACE) Primer.* Available at http://www1.eere.energy.gov/wip/solution center/pdfs/commercial%20pace%20primer%20%28jul%2012%29.pdf. Last accessed 24 September 2012.

US EPA (2011a) *Renewable Portfolio Standards factsheet.* Available at http://www. epa.gov/chp/state-policy/renewable_fs.html. Consulted 21 June 2011.

US EPA (2011b) Personal communication by Ms. Neeharika Naik-Dhungel of the US Environmental Protection Agency (EPA), 3 August 2011.

US Green Buildings Council (USGBC) (2011a) Website. Available at http://www. usgbc.org. Last accessed 12 October 2011.

US Green Buildings Council (USGBC) (2011b) *Case Study Pearl Place, incl. interviews with the Project Architect Manager, sustainability consultant and energy analyst.* Available at http://demo.usgbc.name/projects/pearl-place. Last accessed 12 October 2011.

US Green Buildings Council (USGBC) (2011c) *Case Study Moffett Towers incl. interviews with the project owner representative, the architect and a contractor.* Available at http://demo.usgbc.name/projects/moffett-towers Last accessed 12 October 2011.

US Green Buildings Council (USGBC) (2011d) *Case Study Suzlon One Earth incl. an interview with the Program Director, Synefra E&C Ltd.* Available at http://demo. usgbc.name/projects/suzlon-one-earth. Last accessed 12 October 2011.

Usher, E. (2010) *Engaging the banks in providing end-user financing to the solar water heating sector.* Presented at the Delhi International Renewable Energy Conference (DIREC), 27–29 October 2010. Available at http://www.direc 2010.gov.in/pdf/Engaging%20the%20Banks%20In%20Providing%20End-User%20Financing%20To%20the%20Solar%20Water%20Heating%20Sector.pdf. Last accessed 15 January 2012.

Vethman, P. (2009) *Het financieren van energiebesparing in woningen.* ECN, June 2009.

VfW (2009) *Der Verband fuer Waermelieferung in Zahlen.* Available at www. energiecontracting.de. Last accessed 15 May 2011.

Volkswagen (2010) *Launch of the home power plant – Volkswagen and LichtBlick expand production and sales step by step.* Volkswagen Media Service, Wolfsburg/ Hamburg, 24 November 2010.

Volkswagen (2011) *Lichtblick ZuhauseKraftwerk.* Available at http://www. lichtblick.de/h/Ueberblick_286.php. Last accessed 13 July 2011.

Weill, P. & Vitale, M. R. (2001) *Place to space: migrating to eBusiness models.* Boston, Harvard Business School Press.

Woody, T. (2010) *Loan giants threaten energy-efficiency programs.* The New York Times, June 30, 2010. Available at http://www.nytimes.com/2010/07/01/business/ energy-environment/01solar.html?_r=1&pagewanted=1&hp. Last accessed 11 May 2011.

Woonbond (2011) *Energielabel onderdeel van het woningwaarderingsstelsel.* April 2011.

World Business Council for Sustainable Development (WBCSD) (2007) *Energy efficiency in buildings – business realities and opportunities.* Available at http://www.wbcsd.org/DocRoot/JNHhGVcWoRIIP4p2NaKl/WBCSD_EEB_final.pdf. Last accessed 26 September 2012.

World Business Council for Sustainable Development (WBCSD) (2008) *TEPCO & JFS – energy service companies (ESCO).* Available at http://www.wbcsd.org/Doc Root/U0QJ1vl2BTco1JjB9UlI/TEPCO-ESCO-JFS-Hiro-hospital.pdf. Last accessed 9 September 2011.

World Business Council for Sustainable Development (WBCSD) (2009) *Roadmap for a transformation of energy use in buildings.* Available at http://www.wbcsd.org/ web/eeb-roadmap.htm. Last accessed 10 September 2011.

World Business Council for Sustainable Development (WBCSD) (2010) *Energy efficiency in buildings: transforming the market.* Available at http://www.wbcsd. org/includes/getTarget.asp?type=d&id=MzQyMDU. Last accessed 20 June 2011.

WS Wonen (2010) *Punt van de huur 2010.* WijksteunpuntWonen, April 2010.

Wuppertal Institute for Climate, Environment, Energy et al. (2010) *Change best – Intelligent Energy Europe Project on market development of energy and energy efficiency service companies.* See http://www.changebest.eu/. Last accessed 15 July 2011.

Wüstenhagen, R. & Boehnke, J. (2008) Business models for sustainable energy. In *Perspectives on Radical Changes to Sustainable Consumption and Production (SCP)* Andersen, M. M. & Tukker, A. (eds), Sheffield: Greenleaf.

Index

SWOT (strengths, weaknesses,
opportunities and threats) analysis
8–9, 10, 25; considerations for
building owners/investors 54;
considerations for governments
54–5; discussion and conclusions 54;
energy saving obligations business
models 97–9, *100*; ESCO business
models 41–4, *44*; feed-in
remuneration schemes 52–4, *54*;
green building certification schemes
61–3, *63*; on-bill financing 82–5, *84*;
PACE (Property Assessed Clean
Energy) financing 74–6, *77*;
schemes for increasing rent after
implementing EE measures 68–9, *70*

T-Mobile 144–5
tariffs: in feed-in remuneration schemes
47, 52, 54; on-bill tariff programmes
81, 82, 83, 85
taxation 90–1, 128; 'tax-lien financing'
71
technical barriers to RET deployment
11
tenants 25; and barriers to RET
deployment 12, *12*, 13, 16, **17**,
107; and on-bill financing 79; and
schemes for increasing rent after
implementing EE measures **21**, 23–4,
65–70, *66*, *70*, 113–14, 140–2, *140*,
142
threats *see* SWOT analysis
Tokyo Gas, Japan 142–3
transaction costs, as barrier to RET
deployment 15, 16, **17**, **106**, 110
'triple net lease' 76
Tunisia, PROSOL case study 81, 83,
85, 86, 117, 130, 136–8, *137*, *138*,
139

UK 4; BREEAM ('Building Research
Establishment Environmental
Assessment') standard 56, 57, 58,
62; business models based on energy
saving obligations 95, 96, 96–7, 97,
145–8, *147*, *148*; CERT (Carbon
Emission Reduction Target) 146,
147, *147*, 148; Energy Company
Obligation (ECO) 146, 147; Energy
Efficiency Commitment (EEC) 146;
feed-in remuneration schemes 51;

Green Deal 67, 81, 146; on-bill
financing 81; renewable heat
incentives 23; schemes for increasing
rent after implementing EE measures
67; zero-energy buildings *116*
UNEP 136
University of Michigan, solar thermal
water heating system 3
US: building regulations 59; energy
saving obligations business models
97; Energy Star environmental
standard 56, 58, 121; EPC market
35; FEMP (Federal Energy
Management Program) 38; LEED
('Leadership in Energy and
Environmental Design') standard 55,
56, 57, 58, 121, 150–2, *151*, *152*;
on-bill financing 79, 80, 81, 85;
PACE (Property Assessed Clean
Energy) financing 24, 72–3, 75–7,
79; potential for EE investment in
18; Renewable Portfolio Standards
(RPSs) 104, 115
USGBC (US Green Building Council)
57, 58, 150, *see also* LEED
utilities **112**, 125; business models
based on energy saving obligations
95, 99; and on-bill financing 79, 80,
80, 81, 82, 83, *84*, 85

value chain 45, *45*
ventilation systems: BedZED 4;
cost-effective potential for in
Netherlands *18*
VERDE environmental certification,
Spain 58
Vethman, P. 118
Vitale, M. R. 5, 5, 8
Volkswagen case study 94, 117, 143–4,
144

wall insulation, cost-effective potential
for in Netherlands *18*
WBCSD (World Business Council
for Sustainable Development) 6, 18,
19
weaknesses *see* SWOT analysis
Weill, P. 5, 5, 8
'white certificates' *see* energy saving
obligations, business models based
on
wind turbines; small-scale 2